JUST EARTH

PRAISE FOR *JUST EARTH*

'Tony Juniper, as usual, has called this right. He explores a crucial issue with verve and style. Everyone should read this book.'
George Monbiot, author of *Regenesis* and *The Invisible Doctrine*

'Remarkably well researched, well written and well balanced... nevertheless optimistic about the way forward.'
Richard Wilkinson and Kate Pickett, authors of *The Spirit Level*

'Remarkable, insightful and timely. Juniper sets out an agenda for a just transition and action at all levels.'
Jake Fiennes, author of *Land Healer*

'A compelling read for anyone who cares about all lives on our planet. Tony provides a clear-eyed view of how we got here and what we need to do next.'
Tanya Steele CBE, CEO of WWF-UK

'Based on decades of experience and curated expertise, this book will convert you to the urgent need for a just transition to a better world, where nature and our own species recover together. It also makes clear that we have no other choice. Read it and act.'
Beccy Speight, CEO of RSPB

'At long last! Just Earth is the book that the environmental debate has been missing for decades... it is truthful and robust whilst also being entertaining and hopeful. This is a cracking book, that must set the agenda for the years to come.'
Craig Bennett, CEO of The Wildlife Trusts

'Just Earth powerfully reminds us that social justice and environmental sustainability are indeed two sides of the same precious coin.'
Jonathon Porritt CBE, author of *Hope in Hell*

'A deeply personal account of his own developing understanding that justice for nature cannot be won without justice for people. It's a great read.'
Tom Burke CBE, co-founder of E3G

JUST EARTH

How a fairer world will save the planet

TONY JUNIPER

ADDITIONAL RESEARCH BY LAURA FOX

BLOOMSBURY CONTINUUM
LONDON · OXFORD · NEW YORK · NEW DELHI · SYDNEY

BLOOMSBURY CONTINUUM
Bloomsbury Publishing Plc
50 Bedford Square, London, WC1B 3DP, UK
29 Earlsfort Terrace, Dublin 2, D02 AY28, Ireland

BLOOMSBURY, BLOOMSBURY CONTINUUM and the Diana logo are trademarks of
Bloomsbury Publishing Plc

First published in Great Britain 2025

Copyright © Tony Juniper, 2025
Graph illustrations © Nick Avery

Tony Juniper has asserted his right under the Copyright, Designs and Patents Act, 1988, to be
identified as Author of this work

For legal purposes the Acknowledgements on p. 311 constitute an extension of this copyright page

All rights reserved. No part of this publication may be: i) reproduced or transmitted in any form,
electronic or mechanical, including photocopying, recording or by means of any information
storage or retrieval system without prior permission in writing from the publishers; or ii) used
or reproduced in any way for the training, development or operation of artificial intelligence
(AI) technologies, including generative AI technologies. The rights holders expressly reserve this
publication from the text and data mining exception as per Article 4(3) of the Digital Single
Market Directive (EU) 2019/790

Bloomsbury Publishing Plc does not have any control over, or responsibility for, any third-party
websites referred to or in this book. All internet addresses given in this book were correct at the
time of going to press. The author and publisher regret any inconvenience caused if addresses have
changed or sites have ceased to exist, but can accept no responsibility for any such changes

A catalogue record for this book is available from the British Library

Library of Congress Cataloguing-in-Publication data has been applied for

ISBN: HB: 978-1-3994-1070-0; TPB: 978-1-3994-2600-8; eBook: 978-1-3994-1072-4;
ePDF: 978-1-3994-1067-0

2 4 6 8 10 9 7 5 3 1

Typeset by Deanta Global Publishing Services, Chennai, India
Printed and bound in Great Britain by CPI Group (UK) Ltd, Croydon CR0 4YY

To find out more about our authors and books visit www.bloomsbury.com
and sign up for our newsletters

For product safety related questions contact productsafety@bloomsbury.com

Contents

Preface	vi
1 One Minute to Midnight	1
2 Targets and Treaties	19
3 Never Had It So Good?	42
4 Affluence and Effluents	70
5 Hot House	95
6 Human Nature	120
7 Global Stand-off	146
8 False Choice	172
9 Growth	201
10 Consumerism, Status and Trust	230
11 A Just Transition	253
12 Thrivalism	282
Acknowledgements	311
Image Credits	315
Notes	318
Index	343

Preface

My fascination with the natural world has taken me on a lifelong journey. Many vivid memories have been stowed along the way, and when I sat down to write this book, which is very much about the here and now, I found myself thinking about my childhood. I remember in particular a visit to the Isle of Wight in the summer of 1969, when I was eight years old. I'd gone there with my parents, my two sisters, and an aunt, uncle and cousins, to spend a week at a holiday camp at Freshwater Bay. We didn't have a car, so such trips were rare, and I remember how excited I was, and how determined to make the most of the opportunity to feed my growing passion for nature.

That week, I clambered over cliffs, and nearly fell off one. I explored wild rock pools, catching small fish that hid under rocks for identification. We went on long walks over the chalk hills, following the crest of Tennyson Down to the impressive stacks of white rock known as the Needles that mark the western tip of the island. The flower-rich grasslands were then alive with different kinds of butterflies, most of which until then I had only known from books: clouded yellows, marbled whites, chalkhill blues and more. The orange-and-black-striped caterpillars of cinnabar moths were everywhere, growing fat on the ragwort leaves where the beautiful blue and red adult insects had laid their eggs.

There were whitethroats, skylarks and meadow pipits. Linnets flew between patches of thorny shrub. I sat with my father and peered through my treasured binoculars, a present from the previous

Christmas, watching a kestrel hovering over the grasslands, ready to dive should a vole appear in the open. I caught sight of common lizards darting off the top of anthills when we approached. I learned how to creep up on them to get a better look.

That week became all the more memorable for the history that was being made. The newspaper that my parents read each morning at breakfast was filled with stories about the Apollo 11 mission to the moon, with the climax reached on 20 July when Neil Armstrong and Buzz Aldrin stepped out of their craft to set foot on the lunar surface. That night I lay on the grass outside our little chalet and looked up at the waxing crescent shape above me, imagining what the Earth must look like from up there, from a place where astronauts had to take their own air and water to survive. I thought about the miraculous presence of the birds, plants and insects that I'd seen, and about how lucky they and we were to be in this amazing living place. That week was just one among many experiences that brought me to the writing of this book.

Now in my fifth decade of environmental work, I have seen firsthand many faces of the ecological crisis that is now upon us, and pondered from different perspectives what might be done about it. I have been a lifelong naturalist, spending much time in personal study of the natural world. I've worked as a scientist in global conservation networks, and as a campaigner leading national and international environmental movements. I have worked with primary school children to foster their experiences of nature. I've been a candidate for elected political office, a policy advocate and an adviser to major global companies. I have been an adviser to Prince (now King) Charles, and I am a Fellow with the Cambridge Institute for Sustainability Leadership. I have had the privilege to author many books, and latterly to lead, as its chair, Natural England, the British government's nature agency in England. The roots of all this involvement lie somewhere in my youthful experiences, gazing up at the moon on a long summer night more than half a century ago.

This journey has led me to the conclusion that the huge environmental challenges we face cannot be resolved in isolation, away from the even deeper and bigger questions that they are embedded within, including their social context. This rather important point has of course not gone unnoticed by others, and over the years there have been repeated calls for not only a greener future, but also a fairer one. These twin ideas of 'fairer' and 'greener' have, however, often been rather weakly connected. While most of us get the green bit, the question of whether it really needs to be fair is often overlooked. And anyway, what does fairness mean in a world driven by economic growth?

My conclusion is that if we want to build a secure future, both environmental priorities and social justice must be pursued together. Indeed, so tightly connected are these two challenges that we need to revolutionize the ways we approach development and improving people's lives.

That is the subject of this book, and I was inspired to write it by a chance encounter at a dinner hosted by the London School of Economics in 2008.

I was there with a group of academics, politicians and campaigners to discuss climate change, and what to do about it. A short time before, I'd stepped down as director of Friends of the Earth, where we had recently secured victory in our most significant campaign ever, for the enactment of new laws to cut climate-changing emissions, resulting in what was then the world-leading Climate Change Act 2008. I was seated next to former Swedish environment minister Lena Sommestad. I asked her why Sweden was so much greener than the UK, appearing to be ahead on everything from vehicle standards to energy efficiency, and from sustainable farming to nature protection.

'It's simple,' she said. 'It's because we are so much more equal than you are.'

I had been expecting her to speak about people foraging mushrooms in the forests, wild swimming in big lakes or generally spending a lot of time outdoors. Instead, her straightforward reply underlined how the biggest environmental dilemmas we face are not about pollution, climate change or the destruction of nature, but about society.

Back in 2008, I was already very aware of how the consequences of ecological damage fall unequally across societies, with people of colour, the voiceless and those on lower incomes more likely to be impacted most severely by climate change. They have increased exposure to toxic pollution and less access to high-quality green spaces. In some cases, women suffer more than men, and future generations are disadvantaged too, as they will be dealing with the consequences of choices made now, but without a say in making them. At Friends of the Earth, we'd focused on these dimensions of environmental justice, and highlighted how the disadvantaged groups who suffer the most severe effects of ecological degradation tend to also be the least responsible for causing environmental problems in the first place.

It goes without saying that there are massive inequalities in how much stuff we all use up, and that there is insufficient ecological space for billions more people to increase their consumption to the level now seen across many western societies. This poses a complex conundrum, not least because of how consumption is so fundamentally bound up with the way the world approaches economic development. Growth has become the unchallenged principal means of combatting poverty, even though in recent decades economic growth has also increased inequality, while at the same time driving rocketing demand for natural resources.

Moreover, there is a cultural dimension that means inequality only fuels more consumption. It is linked with a psychological phenomenon called 'status anxiety', which describes the stress and discomfort people feel when they are not keeping up with others who, in material terms, seem to be doing better than they are. The resultant psychological stress leads people to seek out and acquire unnecessary 'positional' goods. Fashionable clothes, luxury cars and expensive gadgets are among the products that are desired, with consequences that include increased debt, erosion of social capital and a weakening of the collective endeavour needed to meet environmental challenges.

I've also observed how, when the issue of inequality arises, it is usually to the detriment of environmental policy. The process of seeking remedies for environmental challenges is routinely blocked, limited or weakened, often (ironically) to protect the interests of these same disadvantaged groups who are most impacted. There are hundreds of examples from around the world of positive environmental proposals being challenged and reversed over concerns about increased expense for people with limited means, especially when a cost-of-living crisis prevails.

At the global level, countries clash and fail to agree on the targets and actions they know are needed to meet environmental goals because of attempts to protect the interests of their less-well-off citizens – who, for example, rely on cheap coal-fuelled power. It's a trend that has also become weaponized, as differences between rich and poor are routinely used – for example, in election campaigns – to advance political ideologies with anti-environmental dimensions, the plight of poor people being held up as a reason not to cut pollution or protect nature.

And yet, these disadvantaged groups rarely have any say in decision-making themselves. One powerful example is the limited political voice possessed by Indigenous societies. There are many instances of such peoples being more focused on sustainability than the pro-development political and economic systems that replaced their traditional ways of living, making their relative lack of power an important ecological issue. While those deciding on policy continue to comprise a narrow social segment, it is unlikely that we will find a route to sustainable development.

All of this merely confirms the size of the challenge and the scale of the constraints that prevent us making the changes needed. Barriers to progress towards environmental solutions not only include the absence of suitable technologies or good policy ideas, they also involve, crucially, matters of equality. This leads me to conclude that greater fairness is a precondition for sustainability. However, it is important to remember that although the disadvantaged are often hit first and hardest, we will *all* be victims of ecological decline. The purpose of *Just Earth* is to prove that the notions of greener and fairer are fundamentally connected, and to set out some of the steps that might help us create a fairer, greener society.

In the pages that follow I present my personal views on the issues at hand, combining my own experiences with a rich array of published literature, as well as the testimony of many experts, through their own words drawn from conversations conducted for this book. The story opens with a look at the world we inhabit today, where environmental pressures are growing as a population of more than eight billion people seeks to expand wealth and prosperity via more economic growth.

I

One Minute to Midnight

We are all broadly aware of how we ended up here, facing unprecedented ecological challenges, but that story is vital to recount in setting the scene for what will follow.

It is, of course, a story that goes way back. Human civilization as we experience it today is the result of a long journey through history. Our modern era, which is the focus of this book, only began around 200 years ago in a small, tranquil hamlet set among fields and saltmarsh. Home to about 25 residents, tucked by the side of the river Tees in north-east England, this quiet corner – surrounded by wetlands, hills and a patchwork of small farms – was to blaze a trail that would change the world. Close to rich seams of coal, along with nearby iron ore deposits in the Cleveland Hills, it was here that an industrial cluster arose. Its name was Middlesbrough and, almost by accident, it would help give birth to the modern world.

By 1841, Middlesbrough was home to over 5,000 people. Abundant natural resources coupled with the rapid rise of new technologies, such as the steam locomotive, led to industrialization. The first port was built, along with workers' homes, foundries and factories burst into life, and a major shipbuilding industry emerged on the south bank of the Tees. On the other side of the river, chemical factories sprang up. New steam-powered ships and railways enabled efficient movement of heavy goods and connected this burgeoning industrial centre with global markets across the fast-expanding British Empire.

By 1889, the population had reached about 80,000. Dense grids of terraced houses began to spread, swallowing up villages listed in the Domesday Book. The broad, wild estuary of the Tees was filled with industrial waste, including slag from furnaces. The once-clear river became grossly polluted, killing nearly everything that lived in it. At night the sky was orange with the fires of industry, while by day smoke spewed from a forest of chimneys, releasing toxic substances into the air and turning buildings black.

The clouds of smoke and smog that lingered over the new industrial cities became heralds of prosperity, for as England's Industrial Revolution burst forth, it generated vast wealth. Money ploughed in by the government, private investors, companies and entrepreneurs returned huge profits. Jobs were created in the new towns and cities, attracting people from the countryside in droves. So compelling was this new industrial idea, so powerful the technology that drove it, that it spread fast and far: first to Europe, then to the United States, and by the late twentieth century to much of Asia. Around the turn of the twenty-first century, the process of industrialization that had taken a century or more in Europe was compressed into a few short decades in China.

Propelled by invention, culture and cooperation, the unprecedented achievements during this period seem to prove irrefutably that our ability to improve the state of people's lives is limitless. Harnessing technology and natural resources, we've found ever more sophisticated ways to meet our needs and desires for improved security, comfort and longevity.

There is a deep paradox, however. The spectacular success of the Industrial Revolution, and of the farming and urbanization revolutions that came with it, is the very reason the future of civilization is now at risk: we developed and expanded our demands without taking our world's life-support systems into account. What is worse, we've known about this paradox for decades.

The warnings were, in the beginning, widely dismissed as being based on incomplete data, pessimistic scenarios and exaggerated projections. But as time went by, the truth became inescapable. Today, it is inarguable that we have a major crisis on our hands, and that this crisis arises from the massive mismatch between our ever-increasing collective demands and the limited capacity of Earth's living systems and finite resources.

Many point to the recent explosion in the number of people on the planet. When modern humans evolved in Africa about 160,000 years ago, the population was small. Indeed, at times in our distant early history there were so few people that by today's standards we would have been classified as an endangered species. Lacking claws, powerful jaws and the ability to outrun predators, we fragile, naked and upright humans relied on social cooperation and the inventiveness of our big brains. Life was dangerous and, mostly, short. Disease, injury, starvation and wild animals all took their toll. But humanity survived nevertheless.

Eventually we colonized every continent except Antarctica, but throughout prehistory our numbers remained tiny. At the start of the Holocene epoch about 11,700 years ago, the global human population was probably around 1.5 million. The simultaneous rise of agriculture in the Middle East, East Asia and the Americas about 10,000 years ago led to more predictable food supplies and the establishment of the first urban areas, which in turn facilitated the rise of a non-farming cohort engaged with administration, writing, engineering, art and religion. The human population began to slowly, but steadily, increase. By the first century AD, the total world population was about 170 million.[1]

It was during the early Industrial Revolution that the population passed one billion for the first time. It doubled to two billion in the 1920s. In 1960, the year I was born, the world population passed three billion, and then four billion in 1973, five billion in 1986, six billion in 1998, seven billion in 2011, and eight billion in 2022. Should I be lucky enough to live to 90, in my lifetime the population may have more than tripled, reaching about 9.8 billion.

The relative biomass of wild animals (basically, their total weight) compared with that of humans and their domesticated creatures (such as sheep, cows, poultry, etc.) is a powerful indicator of this large-scale population shift. Ten thousand years ago, about 1 per cent of the biomass of Earth's air-breathing vertebrates was made up of humans and their animals. Today, about 96 per cent of the air-breathing-vertebrate biomass is comprised of humans and their domesticated livestock, with only about 4 per cent of the total made up of wild creatures.[2]

This rapid population increase arose from laudable changes, ranging from improvements in basic sanitation (reducing fatal illnesses such as cholera and dysentery) to progressive advances in medicine, as well as more comfortable homes, better nutrition, access to education and, for many people, a less physically demanding day-to-day existence. Industrialization, urbanization and the rise of intensive industrial farming were all necessary for a population boom, and they all took off at about the same time 200 years ago in that small English hamlet by the river Tees.

MELT

Rising living standards were an obvious upside of industrialization. The downsides were less apparent at first, but ocean sediments and tiny bubbles of air trapped in polar ice, which are invaluable sources of information about our planet and its atmosphere, tell us that huge shifts were under way. In 1850, the atmospheric carbon dioxide concentration was below 290 parts per million. By about 1910, it had passed 300 parts per million, in 1990 it crossed 350, in 2013 the level went past 400, and in 2022 was at 420 parts per million, the highest level at any time for 800,000 years, and probably for a couple of million years before that too.[3,4]

The pace and scale of change is off the scale compared with anything seen previously during the human era, and that still-thickening blanket of carbon dioxide, combined with methane, nitrous oxide and other heat-trapping gases, is rapidly warming our planet. As we know, the consequences are severe: extreme weather events, the destruction of seasonal patterns and rising sea levels. Greenland now sheds some 270 billion tonnes of melted ice into the ocean each year.[5] It is expected that by 2100, a sea-level rise of about one metre could occur, which will impact the several billion people living in coastal areas and major cities (such as London, New York, Shanghai and Jakarta). Some of the world's most productive farmland will also be lost to rising seas.

Spikes of extreme heat will become increasingly common. Records are already broken, and often smashed, on an increasingly frequent basis. In 2022, England saw 40 degrees centigrade for the first time ever, crushing the previous record set only three years before. The highest global temperature on Earth for 150,000 years was believed to have been reached in 2023.[6,7] Hot and often dry conditions elevate the risk of uncontrollable fires, which have afflicted drought-stricken areas from Australia to Siberia and from the Amazon to California.

The warming of the atmosphere also renders other extreme conditions more pronounced and more likely. Broadcasts from across the world already show us bridges and homes being swept away in flash floods, smashing vital infrastructure and stripping soil from the land. From Germany to Pakistan and from Sierra Leone to Colombia, many countries are suffering the tragic human and economic consequences arising from never-before-seen weather extremes. The question is no longer whether we have a serious challenge on our hands, it is at what point will it become unmanageable?

The web of life that sustains stable conditions on Earth is disrupted by global heating, but it's also unravelled by the degradation of different ecosystems, from coastal wetlands to tropical rainforests to temperate grasslands. Much of this damage stems from how we feed ourselves. For when the world embarked on its industrial path, the population increase and mass migration of people to the towns required a parallel revolution in food production. That too began in England.

Some of England's most productive farmland lies around where I live in Cambridge. The better-drained areas have been cultivated since long before the Roman occupation of Britain, but modern farming originated there during the seventeenth century. It was then that the Norfolk four-course crop rotation system was adopted, which involved growing crops such as turnips and clover to feed animals, and leaving fields fallow for periods to allow the soil to recover. This led to higher crop yields, as well as more meat and dairy. The land available for agriculture was also expanded, through the removal of woodlands, the ploughing of grasslands and the draining of wetlands.

Steam-powered tractors first appeared on British farmland during the 1850s, tearing deep furrows in the soil far more quickly than horse-drawn ploughs had been able to, which was especially useful on hard-to-work land. Industrialized methods led to an increase in the average size of farms, thereby enabling them to benefit from economies of scale. Canals, and then railways, allowed producers to connect with fast-growing urban markets. Land ownership became more concentrated, with successive Acts of Parliament ending centuries-old rights to graze animals. Common land was enclosed as entitlements for collective use were ended, and transferred to private ownership.

Like industrialization, the agricultural revolution wasn't a single event. In the period after the Second World War, the 'Green Revolution' supercharged the process. Governments, international agencies, agricultural companies and investors poured vast sums into land clearance, mechanization, the development of fertilizers and pesticides, as well as irrigation and a further intensification in the selective breeding of plants and animals. Food output soared, leading to lower food prices for urban workers, who became used to having more of their income to spend on other things, including a rising torrent of consumer goods.

The food revolution was a truly global revolution, and (on its own terms) hugely successful, but this success has come at a terrible cost. The elimination of woodlands and grasslands caused habitat loss, while the use of insecticides and herbicides led to the deliberate extermination of much of what had managed to hang on. Our insatiable appetite for land continues today, especially across the remaining frontier lands where largely natural vegetation is still intact – for example, the savannahs that fringe the Sahara Desert, the rainforests of the Amazon Basin and the Cerrado region of central Brazil.

Between 1970 and 2020, the populations of wild birds and mammals (for which we have monitoring data) fell by nearly 70 per cent, while the human population doubled.[8] In 2019, a United Nations (UN) expert study revealed that around one million species (of the eight million or so that presently share the Earth with us) are at risk of extinction, a figure a thousand times larger than what might be expected in the absence of human pressures. The biggest single driving force of this rapid decline in natural diversity is our food system.[9]

Soils, which are the very basis of food security, are another major casualty of our food system. Erosion from agricultural fields is estimated to be between 10 and more than 100 times higher than the soil-formation rate, affecting agricultural yields, in part because of the reduced ability of eroded soil to hold water. Without soil, we won't be able to feed our population, but this soil damage also speeds up climate change, with an estimated 133 billion tonnes of soil organic carbon (SOC, the amount of carbon retained in the soil after the decomposition of the organic content) lost to the atmosphere because of historical damage, which is equivalent to the emissions arising from fossil fuels at around four years of the current annual rate of combustion.[10] And that trend hasn't stopped. Another 27 billion tonnes of SOC is projected to be lost between 2010 and 2050.

Perhaps ironically, the damage that our food system inflicts on the environment is in turn threatening food security and the agricultural industry itself. On top of the potentially catastrophic effects of soil damage, the decline of insects (deliberately wiped out by the use of exotic cocktails of toxic pesticides) is leading to a loss of pollinators, which threatens an annual global crop output worth between US$235 billion and US$577 billion.[11]

In addition to being a result of how we feed ourselves, the loss of diversity has also arisen from the direct exploitation of wildlife, whether in the form of catching wild fish, hunting rare species for traditional medicines or, for example, cutting down natural forests for wood. These all disrupt habitats that are already being fragmented and sliced into ever smaller patches by agriculture, urbanization and infrastructure. The disruption caused by invasive non-native species is also enormous, and as the world warms, the impacts of climate change often exacerbate the effects of the other pressures bearing down on the health of the natural world.

The extraction of natural resources used to feed our material consumption is another rapacious driver of ecological decline and climate change. In 2020, the minerals, metal ores, fossil fuels and biomass (including wood) extracted exceeded 100 billion tonnes for the first time ever.[12] That is the amount required to create all the new buildings, roads, fertilizers, aircraft, cars, computers, chairs, trains, windows, paint, paper, phones and all the rest of the paraphernalia that surrounds the world's eight billion people. Behind that number lies a vast and expanding effort to extract natural resources from beneath the ground and from ecosystems, including forests and the ocean.

Most consumption, however, ultimately turns into waste, with resources and products ending up in landfills or incinerators, or polluting the environment, including the millions of tonnes of plastic that flow into the ocean each year.[13] This is now, literally, a planet-wide phenomenon. Microplastic fragments are ubiquitous, plastic bags have been found in the Mariana Trench at the bottom of the deepest ocean, and there's even a waste dump located 8.6 kilometres above sea level on Mount Everest. Litter is the most visible result of our consumption, but the environmental consequences begin much earlier, at the point of extraction, and increase when raw materials (such as fuel for engines) are processed, refined and used. Steps towards the creation of a circular economy, whereby resources can be recovered and reused, have, however, been modest. For example, 460 million tonnes of plastic is produced each year, and yet only around 9 per cent of it is recycled.[13]

Even though we know that ever-increasing consumption leaves in its wake scarred landscapes, polluted atmospheres and the onset of mass extinction, the world's demand for raw materials has only ramped up and shows no signs of slowing down or stopping.

On the contrary, the plan in countries right around the planet is to at least maintain, or, more usually to increase, economic growth, which, if historical patterns are repeated, will mean even more demand for resources, energy and land. With growth policies that pay insufficient regard to the long-term implications for people and the environmental assets that sustain them, attempts to reduce our collective impact rarely succeed at the scale needed.

Many people still regard environmental degradation as an unavoidable consequence of development. It's true that the economic thinking we have embraced has generated wealth and supported a rising population, and the assumption by governments and in company boardrooms around the world has been that the resultant ecological damage is a regrettable, but nonetheless inevitable, part of a process that permits social progress via economic growth.

This view is, however, deeply flawed. A 2024 study by the Potsdam Institute published in the journal *Nature* warned that climate-change impacts could lead to much more economic damage than previously estimated.[14] Even if carbon emissions were cut rapidly right away, the study showed that a cut of global income by nearly a fifth could occur by 2050. What is also very important to know is that this loss of economic value is about six times as much as it would cost to decarbonize the economy to meet the two-degree warming limit set out in international agreements. More on that in the next chapter, but in the meantime the important point from this study is that it would be much cheaper and economically more rational to cut emissions quickly than to carry on growing the economy as we historically have done. A case in point as to how such economic numbers might manifest in real world events burst on to TV screens in January 2025, when devastating fires coming in the wake of eight months without rain tore through parts of Los Angeles in California. Early estimates placed the costs of these vast conflagrations between tens to hundreds of billions of dollars. And on top of climate change, the continuing degradation of nature has economic security implications too.

Food and water security is dependent on functioning natural systems while access to nature sustains public health. The most cost-effective strategies for adapting to climate change are often nature-based solutions, such as more natural coasts that can cushion communities from the effects of sea-level rise and storms. It's hard to accurately assess its economic value, but one of the most thorough estimates posits that the total annual contribution of nature to the economy is considerably greater than the global gross domestic product (GDP).[15] GDP is the headline economic figure that countries across the world seek to grow (more on that later), so the fact that nature underpins it is a very important and fundamental point.

So much for the idea that promoting economic growth and meeting environmental goals are different things. They are not. As I have written elsewhere, including in my 2013 book *What Has Nature Ever Done For Us?*, the truth is that we inhabit a linked system, with nature sustaining the economy. Every now and again we receive a vivid reminder that the two are connected, and that disruptions to nature can have a major impact on the human world. This includes the emergence of novel diseases that have caused massive economic and social disruption.

During recent decades numerous pathogens have jumped into the human world, causing death and mayhem in their wake. AIDS, Ebola, SARS, Nipah virus, COVID-19 and many other diseases came from animals to humans with devastating effects, not only for the individuals who were infected and sometimes died, but also for society. The risk of diseases spreading to the human population is ever-present, but this risk is heightened through our changing relationships with the rest of life on the planet.[16] AIDS came from the hunting of apes to supply the growing trade in bushmeat. Ebola is linked with deforestation and people encountering wild species in what were once remote areas. Nipah virus, a deadly pathogen that emerged in Malaysia, jumped from bats to pigs and then to people. This deadly virus was fortunately contained before it spread too widely, but it could have caused a major emergency.

One virus that we failed to contain was COVID-19, which probably came from horseshoe bats and possibly passed to another animal in captivity – maybe illegally traded pangolins (a rare species of anteater) – before mutating and getting into humans.[17]

On top of this are the effects of climate change and the extent to which pathogens might expand into new geographies as the world warms. One recent estimate suggests that changes to the natural world might result in over 10,000 more viruses coming into closer contact with people during the coming decades.[18]

THE GREAT ACCELERATION

One person well placed to describe our present predicament is Johan Rockström, Professor in Global Sustainability at the University of Stockholm (as well as being Professor in Earth System Science at the University of Potsdam), who leads the influential planetary boundary science work at the Stockholm Resilience Centre. Johan is a brilliant scientist with whom I am lucky enough to have worked as a fellow trustee at the rainforest charity Cool Earth. He told me that we have modified our planet to the point of initiating a new geological epoch: the Anthropocene, or the era of humans.

'We are now in an entirely human-created geological epoch,' he told me. 'We even have a date for the entry of this, in the 1950s, which is the take-off point of all these exponentially rising hockey sticks of increasing human pressure on the system.' The 'hockey sticks' he refers to are moments when more or less flat-lining trends suddenly take on a rapid upswing, like the shape of a hockey stick. They are evident in the post-war population explosion, the expanding size of the global economy, the rising concentration of carbon dioxide in the atmosphere and the increasing demand for natural resources. Rockström and his Stockholm colleagues have dubbed the combined rise of all those curves (and many others) 'the Great Acceleration'.

Rockström calls it 'an extraordinary system shift'. 'We're leaving one geological epoch, the Holocene, which is the only state of the planet we know for certain can support humanity, and then we're now in the Anthropocene,' he says, pointing out how we have reached a point of major change. 'We are 70 years into the Anthropocene and we are starting to see how in the first decades – the 1950s, 1960s, 1970s, perhaps even in the 1980s – the Earth system clearly had a very significant bio-geophysical resilience to cope with the stress and the pressures caused by the Great Acceleration.'

The natural world has absorbed more than half of the emissions released since the start of the Industrial Revolution, initially protecting us from the effects of pollution that would otherwise have become apparent more quickly. Rockström, however, believes that during the past few decades we reached a 'saturation point'. 'We are hitting the ceiling of the hardwired biophysical processes and systems that keep the planet stable.' And not only have we arrived at the limit of what our planet can do to compensate for our demands, Rockström and his teams have concluded that we are at the threshold of triggering a series of 'tipping points', beyond which the whole Earth system will become unstable.

Rockström argues that 40 years of research reveal that major change is at hand. 'The Earth is an interconnected, complex self-regulating system,' he says, but he believes that this system is at serious risk of destabilization, with the evidence for that also very compelling. 'We're talking about the Greenland ice sheet, we're talking about the Atlantic overturning of heat in the ocean, we're talking about Antarctica, we're talking about the big rainforests. We're talking about the permafrost in the boreal region, we're talking about Earth's tropical coral reef systems.'

All these major components influence conditions across the entire world. The total disintegration of the Greenland ice sheet would add seven metres to the global sea level; the circulation in the Atlantic determines climatic conditions across vast swathes of densely populated territory; the conditions in Antarctica influence both global climate and global sea levels; the Amazon rainforest holds billions of tonnes of carbon and a vast wildlife diversity, while trapped within the frozen fabric of the permafrost lie billions of tonnes of not only carbon, but also the powerful greenhouse gas methane, and should that defrost, then the pace of global heating would accelerate at a whole new level. The loss of the coral reefs would contribute to, among other things, the mass extinction of many marine species.

THE GREAT ACCELERATION

FUNDAMENTAL PRESSURES

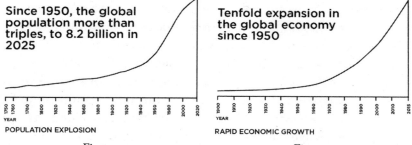

Figure 1 — POPULATION EXPLOSION: Since 1950, the global population more than triples, to 8.2 billion in 2025

Figure 2 — RAPID ECONOMIC GROWTH: Tenfold expansion in the global economy since 1950

EXPANDING DEMAND

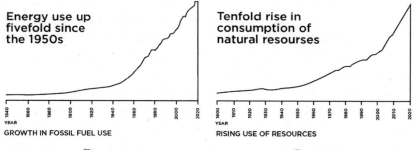

Figure 3 — GROWTH IN FOSSIL FUEL USE: Energy use up fivefold since the 1950s

Figure 4 — RISING USE OF RESOURCES: Tenfold rise in consumption of natural resources

CONSEQUENCES

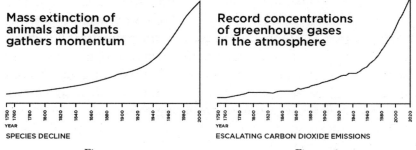

Figure 5 — SPECIES DECLINE: Mass extinction of animals and plants gathers momentum

Figure 6 — ESCALATING CARBON DIOXIDE EMISSIONS: Record concentrations of greenhouse gases in the atmosphere

All this leads Rockström to call our predicament a 'dire situation'. He adds that we cannot rule out the distinct possibility that thresholds have already been crossed, to the point that some of these tipping points might have already been activated. The implications are profound, for without a stable, productive and healthy natural world, vital functions that enable the Earth to function as it does will cease to operate as they do now.

'Without biodiversity, no food system,' he says, 'without biodiversity, no ability of sequestering carbon and no ability to regenerate materials, no ability to regenerate oxygen-breathable air, fresh-water quality, so losing biodiversity poses a fundamental threat to the live-ability on Earth.'

No wonder Rockström is one among a growing number of experts who call for a 'revolution pace' in decarbonizing energy, transport and land use, in creating a circular economy to replace our non-sustainable linear economy, and in shifting to regenerative farming and restoring ecosystems at scale. This fundamental shift is far more than an optional environmental programme, it is a prerequisite for the survival of human civilization. We can't limit global heating without protecting and restoring nature, and it will be impossible to halt the loss of species and the degradation of ecosystems if rapid climate change continues with emissions levels remaining unchecked.

It's a challenge that will require interconnected action. Professor Sir Bob Watson describes it as a 'triple crisis of climate, biodiversity loss and pollution' which 'all interact with each other'. Among other things, Watson is a former scientific advisor to the British government, former Chair of the Intergovernmental Panel on Climate Change (IPCC) and Chair of the Intergovernmental Science-Policy Platform on Biodiversity and Ecosystem Services (IPBES). I am fortunate to have interacted with Bob in several of his major roles and have been impressed by his rigour and devotion to evidence as well as his analytic ability to see the connections between different aspects of our changing world.

When I spoke to him for this book he pointed out how, in seeking remedies to multiple challenges, 'you have to look for a common solution for all of them simultaneously'. He is one among many leading experts who have reached the conclusion that climate change and ecosystem degradation are not in the end environmental issues. 'I actually think we should be viewing them as economic issues, development issues, security, social, moral and ethical issues, because all of them undermine economic development.'

His decades of work lead him, like Rockström, to conclude that we must now act with a real sense of emergency. 'All of this crap on climate change about net zero by mid-century is missing the key point,' he says. 'That's fine. We do need net zero by mid-century. But what we need is action between now and 2030 to reduce the emissions for climate change anyway, by about 50 per cent. And the same on biodiversity, to reduce pollution, reduce land degradation. It will be the next few years that will make all the difference. As far as I'm concerned biodiversity loss is equally as important as addressing climate change. I think if we indeed do continue to destroy biodiversity it's going to directly and indirectly affect people's everyday lives.'

Watson's estimate of the pace of action needed is based on the IPCC's scientific findings, setting out the scale of emissions reduction required to avoid the worst consequences of climate change. This comes down to limiting the average temperature increase globally to below two degrees centigrade, and as close as possible to 1.5 degrees centigrade, by 2100, compared with levels in the pre-industrial period (taken to be pre-1850).[19] This in turn leads to the conclusion that total world emissions must reach zero by 2050, allowing only for carbon dioxide releases that will be recaptured, including by ecosystems (such as newly expanding forests and recovering soils) or different technologies, such as those which capture carbon. The longer we put off the decisive action needed, the more likely it is that we will need to go beyond the goal of net zero, and into a programme of negative emissions, in order to recapture a large quantity of the carbon dioxide that has already been released.

It is not a challenge that can be left until 2050. Our programme must start right away, not only halving emissions by 2030 compared with 2020, but doing the same again by 2040 and getting to zero by 2050. Going beyond 1.5 to two, three, four or even five degrees of global warming will lead to progressively more disastrous impacts. The longer decisive action is delayed, the deeper the peril and the greater the economic costs. One projection from the climate science that underlines the point is the estimate that at 1.5 degrees of warming the Earth will 'only' lose 70 to 90 per cent of its shallow water coral reefs, whereas at two degrees of global heating this ecosystem will be nearly completely wiped out.[20] It is these tight and fundamental connections between global heating and the state of the planet's biodiversity that have put the Earth's habitability in such peril.

On the threat to biodiversity, Watson says that it is vital to address the underlying drivers of the problem. 'Stop deforestation, stop converting the grasslands, stop converting the mangrove swamps. Stop over-exploiting plants and animals. Climate change must stop, otherwise it will become the greatest driver of biodiversity loss. Stop pollution, stop invasive alien species. In reality, unless you get to grips with those drivers, all of the other stuff is irrelevant.'

We are living through the last moments before it will be too late to avoid up to several degrees of global warming and the loss of a high proportion of our planet's web of life via a mass-extinction event. Is it too strong to say that we are at 'one minute to midnight'? I don't believe so.

In May 2024 I visited Bradgate Park in Leicestershire, where I learned about archaeological investigations into the traces left behind by some of the first hunter-gatherer people to arrive in Britain from Europe towards the end of the last Ice Age. This place was an important inspiration for the biologist and natural historian Sir David Attenborough when he was a child. It remains a source of fascination today, and I went there to declare it as a new National Nature Reserve.

I visited the mouth of a narrow gorge that opens onto the floodplain of a small river that in the distant past would have been a good place for hunting, and that is where stone tools and flint shards were found. These faint traces left by the first Britons have been dated to about 14,500 years old.

It is hard to imagine just how long ago that was, or to put it into our modern context, but if you divide that time span into decades, you get about the same number as there are minutes in a day: 1,440. During the first four hours or so of such an imaginary day, the last Ice Age came to its end, and the Holocene epoch began. It was during this unusually stable post-glacial era that the rise of civilization took off, eventually leading to us humans being an omnipotent global presence in the twenty-first century.

On this time scale, the Romans came to Britain at about 10 p.m. and the Iron Age began a little over an hour before that. The British Bronze Age started at about 4 p.m. and the Neolithic, with its first settled farming started at about 11.30 a.m., with the subsequent construction of Stonehenge at around 3 p.m. All human activity before the Neolithic, and since the start of the Holocene, was in a hunter-gatherer economy. The decade we are in now, on this scale, is thus one minute before midnight. This last minute must mark a turning point, setting in motion reversals to the trends that began with the Industrial Revolution in England, and which started only about 20 minutes earlier, and that took off as a driver of profound global change with the onset of the Great Acceleration, about seven minutes before this final critical minute.

In 2023, the IPCC, in its most recent assessment, pointed out that 'the choices and actions implemented in this decade will have impacts now and for thousands of years'.[21] We have no reason to doubt that we dwell in a moment of great consequence, and that the choices we make now will be of huge significance. Those choices could be positive, setting in place a secure collective destiny for humankind and the rest of life on Earth for millennia to come.

It is not a modest programme of change that is needed, however. The scale of what must be achieved is as significant as the great shift that revolutionized the world in the nineteenth century, when civilization embarked on its present path. We need to reverse the steeply rising curves that expose the damaging aspects of the Great

Acceleration, while continuing to improve social well-being. At one minute to midnight, facing the connected crises of global heating, nature decline, resource depletion and toxic pollution, we have run out of time.

It's not as if we didn't know what was going on. For decades, we've had repeated warnings and have set multiple targets to guide action. The real question is, are we making any progress?

2

Targets and Treaties

Environmental campaigners who've been in this work for as long as I have wonder how it is that, after more than thirty years of setting international targets, we're still heading in the wrong direction. Yes, the necessary changes are huge and the process complex (spanning decades, embracing all countries, touching all sectors, and at heart requiring shifts in economic ideas and social systems), but surely we should be on the road to environmental recovery by now? You, the reader, will remember a blur of headlines about COP agreements on limiting climate change, headline-grabbing targets and world leaders shaking hands. Yet we are still so far from where we need to be. Global emissions are rising, deforestation remains rampant and the mass extinction of species is just around the corner. From what I've seen, there are several major factors holding back progress. To discover what's going wrong, it's essential to understand the history of the agreements and targets we fought so hard for.

On 22 October 1990, I reported for duty at Friends of the Earth's London headquarters as the new leader of the organization's tropical rainforests campaign. I came following a stint at BirdLife International, where I had been responsible for work to prevent the extinction of the world's rarest parrot species. Many of these birds lived in tropical forests, so I arrived with some knowledge of what was happening in those parts of the world, where pressures arising from farming, the timber industry, mining and major dams were among the reasons why such ecosystems were being degraded and destroyed. My new role, though, was less about ecological science and more about politics. At BirdLife, our efforts were focused on forensic field science and data, but Friends of the Earth embraced not only science but its application to questions of policy, ethics and rights – and, to a large extent, sought change based on an analysis of where political power and influence lay.

The change in office premises rather summed up my career shift. BirdLife (then still called the International Council for Bird Preservation, or ICBP) was a science-led global network based in a rambling Victorian house in the leafy Cambridge suburb of Girton, populated by expert ornithologists. Friends of the Earth, by contrast, was headquartered in a rather run-down part of the City of London, where the business was all about campaigning. More than once a stock photograph of the street the offices were on (replete with broken windows and a burned-out car) was used by national newspapers to illustrate inner-city decline in England. Inside, warrens of packed metal-framed shelving groaned with documents, banners, boxes of leaflets, stickers, placards, posters and other campaign paraphernalia. It felt gritty, like a front line for change – and it was.

For the most part, the political dimensions of our campaign work focused on the question of what countries were prepared to do to reverse pollution, waste and the destruction of nature. During the 1980s, the public in many nations, including the UK, had started demanding more environmental action from their politicians. This shift came in the wake of grim news and expert reviews detailing what was happening to the environment; the hole in the Earth's protective ozone shield, in particular, generated huge concern. So too did the damage caused by acid rain. The 1989 grounding of the oil tanker *Exxon Valdez* in Prince William Sound in Alaska spilled 37,000 tonnes of crude oil, killing a vast number of wild creatures and sparking international condemnation. By the time I arrived at Friends of the Earth, the range of issues on the table was daunting in both breadth and scale.

We were concerned by the toxic pollution wrecking rivers, by the particulates and noxious chemicals in the air causing a range of health problems, and by the destruction of wildlife habitats across the world, including the loss of the tropical rainforests that I had arrived to do something about. We had people working on everything from waste and recycling to transport and land use, as well as a team dedicated to energy policy, including the massive potential for energy efficiency via simple methods such as insulation and draught-proofing.

To make positive inroads, we had some tried-and-tested methods to generate public demand for change and thereby place pressure on governments and companies. We conducted research and gave the findings to the media, in the hope that this would increase awareness. We organized letter-writing drives, marches and direct actions to shine the spotlight on the right places at key moments. We also encouraged individuals to help by saving energy: cycling rather than driving, and recycling bottles, paper and cans. In addition to our 120 or so staff, Friends of the Earth had a network of about 200 local volunteer groups and about 60 national Friends of the Earth organizations all around the world. The energy in the organization was intense.

Anyone who walked through the door to work there was soon swept along with a sense of excitement and possibility. The organization remains on the front lines today, although since then the environmental field has exploded, with new campaign groups, major companies going green, governments setting new targets, and direct-action networks now all making for a more active and crowded field.

Looking back, it was a golden time for environmental groups, with increasingly engaged ministers, rising public support and informed media coverage. Protests, news stories and letter-writing campaigns were often enough to get the attention of politicians and company executives, and if there was good evidence that action was needed, a response would often follow. In 1990, pressure from Friends of the Earth and others led to the publication of a Department of the Environment White

Paper called *This Common Inheritance*, which was presented by Chris Patten, then Environment Secretary in Margaret Thatcher's cabinet. It was a landmark document, setting out a wide range of actions to achieve better environmental outcomes in the UK.[1] It was so important that our Friends of the Earth campaign team stayed up all night to fully analyse every sentence of the new policy to assess its value, before making our views known. This rise in political interest was bolstered by a range of increasingly urgent warnings from expert bodies, both on the global stage and closer to home.

OUR COMMON FUTURE

If I were to name one moment that heralded the modern period of environmental action – when the environment was solidly on the public radar and truly agenda-setting initiatives were in play – it would be March 1987, with the publication of the Brundtland Commission's report, three years before I joined Friends of the Earth. The commission was led by former Norwegian prime minister Gro Harlem Brundtland, and its report was called *Our Common Future*.[2] I was travelling in Wales with my friend Sue (now my wife of 35 years) when it was announced. High-level calls to action were broadcast on the BBC radio news bulletins as we drove between the wonderful oak woods, dunes and marshes of south-west Wales.

Looking back on that report today, it is striking how closely interwoven the environmental and social challenges were seen to be. If there was to be a secure future, the natural systems which sustained people would need to be sustained as well, it argued. The idea of sustainable development was presented as an integrated challenge, whereby the needs of people would be best served through looking after the environment. Fast forward to the mid-2020s, and the political discussion has in some respects lurched backwards, with sustaining nature and economic development often presented as incompatible choices, rather than, as the Brundtland report saw them, two sides of the same coin.

Our Common Future soon became a touchstone for environmentalists, identifying how continued environmental damage would in the end threaten social progress. The ideas it set out galvanized debate and sparked preparations for a major UN summit hosted by Brazil in Rio de Janeiro in June 1992. Although the summit was called the United Nations Conference on Environment and Development (UNCED), it was to become popularly known as the Earth Summit.

The timing of the Earth Summit was set to mark the 20-year anniversary of the first major global environmental gathering, which took place in Stockholm in 1972. That original meeting marked a period of growing awareness after NASA's Apollo 8 space mission returned with images of a fragile blue pearl suspended in the darkness of space, rising above the horizon of our dead Moon. The photograph *Earthrise*, captured from the window of the command module of Apollo 8 in December 1968, presented people with a scene never before witnessed, allowing them to look back at our world from far outside its atmosphere. That alone delivered a powerful message, inspiring the first waves of organized environmental activism, during which both Friends of the Earth and Greenpeace were founded.

The declaration negotiated in Stockholm set the tone for decades to come, uniting as it did questions of human rights and economic development with emerging environmental priorities. Crucially, it was backed by an action plan and several practical steps. National governments agreed to appoint environment ministers (though a few countries, including the UK, had already done so), the United Nations Environment Programme (UNEP) was set up and World Environment Day (marked on 5 June) was created.

At the time of the Stockholm conference, I was 11 years old and already a fully fledged nature-nut. But I was only dimly aware of the significance of the meeting that popped up on the news from time to time during that summer of 1972. Twenty years later, however, I certainly was engaged in what was happening, appearing on the news myself to draw attention to the need for a successful summit in Rio. I'd been involved with international negotiations through my work with BirdLife International, making the case for stronger regulation in the trade of very rare parrots, but the process leading to Rio was my first experience of working to influence negotiations taking place under the auspices of the UN.

In the early 1990s, the media was much more interested in environmental questions than it had been in the early 1970s, and a steady stream of stories came day after day from a group of specialist correspondents. Among them was journalist Geoffrey Lean, one of the world's first environment correspondents. Lean had a sixth sense for news and understood very well what would get past his editor and into the paper. With detailed and broad knowledge, highly tuned political antennae, endless energy and an unswerving determination to nail stories, he was a one-man dynamo for expanding public awareness. Today, he is possibly the world's longest-serving environmental journalist, and by the time of the Earth Summit he was already a veteran with more than two decades' experience, who had for some years been environment correspondent with the *Observer*.

I asked him about his impressions from that earlier time, and he reminded me of the sense of optimism in the run-up to Rio, not least fuelled by the global environmental progress made during the 1980s on the matter of ozone depletion. Whereas climate change dominates environmental discussions today, during the second half of the 1980s it was the 'hole in the sky' that most prominently drove headlines.

When I came to work at BirdLife International I met with Joe Farman, a modest and unassuming scientist who then worked down the road with the British Antarctic Survey. Using old instruments kept at the Halley Research Station located on the Brunt Ice Shelf, Farman had discovered the ozone 'hole' over Antarctica, which ultimately galvanized a global political process that culminated in the Montreal Protocol, signed in September 1987. This agreement, based on the previously negotiated Vienna Convention, was to be regarded as the most successful international environmental agreement in history, given its impact on phasing out the pollution that was depleting the Earth's protective shield of high-altitude ozone.

'One really important thing about Rio was how it came in the wake of Montreal,' recalls Lean. He credits Egyptian diplomat Mostafa Tolba, then head of UNEP, as the person who enabled the breakthrough. 'He was an absolute genius at getting negotiations completed. He used to leave people in locked rooms and not let them out until they'd finished, that sort of thing.' The Montreal Protocol was so tightly negotiated and nuanced that Lean recalls how Tolba didn't even want it translated from English into the other five official UN languages, for fear that it would spoil the delicately agreed text and the whole thing would unravel. It was, however, immensely successful.

The mood before Rio, Lean recalled, was very much influenced by the triumph of Montreal, with some seeing the prospects for a strong deal on climate change coming in the wake of the one thrashed out to deal with ozone depletion. 'I was certainly very hopeful that a deal could be done. The first world was getting worried about climate change and the developing world was worried about development. I could see the outlines of an agreement whereby the developing countries undertook to join developed ones in acting on climate in return for something substantial on development from the rich countries.' Such was the hope.

THE ROAD TO RIO

A series of four preparatory meetings paved the way for the Earth Summit. These were basically opportunities for countries to come together to prepare items on the agenda. I attended all except the first and witnessed at close quarters the discussions (the last of which took five weeks) on the different subjects that would be prepared for final negotiations in Rio. The talks were streamed into three broad themes. Working Group One looked at the atmosphere, forests and biodiversity; Working Group Two discussed oceans, freshwater and waste; and Working Group Three took on legal and institutional arrangements.

It was complicated and fraught. My colleagues and I in the Friends of the Earth International rainforest team set out to influence the negotiations in Working Group One. With hard-to-follow talks taking place for the most part behind closed doors, the opportunities for generating external pressure via the media and public mobilization were limited, so we had to focus on lobbying the negotiators in person. Our big call was for a new, global Forests Convention, which would be a legal agreement on the conservation and sustainable use of the world's forests.

The talks took place in the Palais de Nations in Geneva, which had been commissioned by the League of Nations during the 1920s. This imposing art deco building was the League's crown jewel, containing dozens of meeting rooms that have over the years hosted thousands of diplomatic meetings. It was outside some of those negotiating rooms that a handful of colleagues and I sought to intercept delegates with ideas for wording to include in draft agreements. However, except for France, Canada and one or two others, there was little support for a Forests Convention. On the contrary, there was active hostility towards it, especially from the developing countries, whose representatives were far from convinced that it would be a good idea to take on binding commitments about the future of their forests.

Greater progress was made on other subjects, including negotiations towards three conventions in relation to desertification, biological diversity and climate change. The process was difficult and bumpy, but the emerging science sustained progress. For example, the IPCC's first assessment report had been published in August 1990 and created a new scientific frame of reference for the talks, and also a benchmark against which, on that subject at least, campaigners might make judgements about the success of the Earth Summit.

The seriousness of the challenge was becoming increasingly stark. At the Rio Earth Summit, a group of 1,700 leading scientists published what they called a 'World Scientists' Warning to Humanity'.[3] 'Human beings and the natural world are on a collision course,' they said, claiming that if the direction of travel was not checked, the living world 'will be unable to sustain life in the manner that we know'. 'Fundamental changes are urgent if we are to avoid the collision our present course will bring about,' these scientists warned. That was in June 1992.

The impending collision was considered by political leaders to be a real danger. At the time, and ever since, it was the discussion about climate change that attracted the most attention, and the talks in Rio led to the agreement of the United Nations Framework Convention on Climate Change (UNFCCC). This agreement was only thrashed out during the dying moments of the summit. Various goals, ambitions and targets were adopted, including the overarching objective of the 'stabilization of greenhouse gas concentrations in the atmosphere at a level that would prevent dangerous anthropogenic interference with the climate system'.[4]

The new convention went on to say that this objective should 'be achieved within a time frame sufficient to allow ecosystems to adapt naturally to climate change, to ensure that food production is not threatened and to enable economic development to proceed in a sustainable manner'. As a first step, the richer countries and the emerging economies of Eastern Europe agreed to stabilize their emissions at 1990 levels by 2000. This was by any current measure a very modest commitment, although at the time, and considering that all of them were expecting to expand their economies during this period, it was a significant undertaking.

Another of the UN's Rio agreements became the principal vehicle for dealing with the loss of species and habitats at the global level: the Convention on Biological Diversity (commonly referred to as the CBD).[5] Getting both agreements over the line was far from straightforward, and in some ways the CBD came even closer to derailment. One reason it nearly fell apart was because the USA refused to sign it (although it did begrudgingly permit an American signature on the climate treaty). President George Bush senior's words at the summit have gone down in history: 'The American way of life is not up for negotiation. Period.' At Friends of the Earth, we were worried that not only would we fail to secure an agreement on forests, we would also fail to get one on biodiversity.

We stayed up through the night to monitor what the negotiators were doing, looking for opportunities to encourage the adoption of a strong treaty. Geoffrey Lean recalls the drama that unfolded as the Rio summit opened: 'The Americans came out against the Biodiversity Convention just literally at the last moment.' Lean told me how a fellow reporter got the story and helped to raise the profile of the jeopardy at hand. 'Greg Neale picked it up and ran it in the *Sunday Telegraph* and then got on the plane and told the other journalists about it on the way. As soon as the plane landed they went and grilled the British delegation, who seemed about to cave in to US pressure, and filed stories that made the splash or page one in many papers. There was a hell of a battle and we were lucky that it was a very slack news week and so for days what Britain was going to do about the Biodiversity Convention was on the front pages. That helped a huge amount.'

So it was that most countries, including the UK, in the end signed the CBD and undertook to pursue its three big aims. The first was the conservation of biological diversity, the second was about the sustainable use of the wildlife and ecosystems (such as fish stocks and forests) that comprise it, and the third was focused on sharing the benefits gained from the utilization of genetic resources, such as those arising from patents on new drugs derived from wild species.

EMBEDDING COMMITMENTS

In the wake of these landmark agreements, the first Conference of the Parties (COP) of the Climate Change Convention was organized. COP1 was hosted by Germany in 1995, in recently reunified Berlin. One of its main tasks was to review the adequacy of the commitments entered into in Rio, and to determine if more needed to be done.

By the mid-1990s, the science of climate change was reaching ever more troubling conclusions and it was apparent that the wealthier countries' commitment to stabilize emissions at 1990 levels was far from adequate. More, much more, was needed. Thus was born the Berlin Mandate, which was an agreement among countries to open negotiations towards a new set of more demanding goals to be taken on by the developed countries. The Berlin Mandate kick-started discussions that would conclude at COP3, two years later, in Kyoto, Japan. Between the Berlin COP and Kyoto, the IPCC published its Second Assessment Report, which moved the dial of scientific certainty a further notch forward with the conclusion that 'the balance of evidence suggests that there is a discernible human influence on global climate'.[6] That was in 1995.

The Kyoto COP was the scene of further political and diplomatic dramas. I was there with a team from Friends of the Earth International, urging the adoption of science-based targets. The summit, which took place during an unusually warm December, saw a showdown when I led campaigners into battle against fossil-fuel companies, inviting conference delegates to vote for who they believed was the most toxic and corrosive fossil fuel entity at the talks. It wasn't only the environmental groups that were worried about climate change. Powerful business interests had also paid increasing attention to what was happening, especially following the agreement reached at the Rio Convention in 1992, investing vast resources in pushing back against agreements to reduce emissions. We issued thousands of ballot papers listing those who we believed were contenders. The winner of what we dubbed 'The Scorched Earth Award' was the Global Climate Coalition, a US-based association of oil, gas, coal and automotive companies that was working to derail prospects for an effective new agreement.

After we'd counted several thousand votes, I invited the British former Secretary of the State for the Environment John Gummer to present the award. Later to become Lord Deben and chair of the UK's official Climate Change Committee, John is a right-of-centre advocate for action on climate change who, having served in the Thatcher cabinet, was a perfect foil for the climate deniers. He unveiled the results at a packed press conference, to which none of the companies in the running had, perhaps unsurprisingly, sent a representative. It generated international media coverage and, alongside other tactical plays, helped to sustain some sense of direction at the talks.

Such was the intensity of the campaign we ran in Kyoto that after the first week of talks I collapsed and ended up in hospital on a drip, suffering from exhaustion and dehydration. I rested for the weekend and was back in action by the start of the second week. Despite the best efforts of those with fossil fuel interests and other climate-change deniers, including countries such as Saudi Arabia, to wreck the negotiations, an agreement was reached. It was called the Kyoto Protocol, and it became a subsidiary deal under the original commitment agreed upon in Rio five years previously. Now, developed nations promised to undertake actual cuts in emissions, rather than just stabilizing against where they were in 1990. The Friends of the Earth team complained that it was too little compared with what the science at the time said was needed, but it did mark progress. Most importantly, it was a significant defeat for the fossil fuel interests.

The Kyoto Protocol encompassed not only new obligations to reduce greenhouse gas emissions in better-off countries, but also mechanisms to assist in cooperation and flexibility in how the emissions cuts were to be achieved. This embraced the idea of emissions trading that the USA insisted on, rather than carbon taxes, favoured by the EU. Emissions trading was a market mechanism that enabled action in one place to be counted as an offset in another. For example, one country might plant trees to suck up carbon and another might buy carbon credits created by that to count towards its own emissions reductions. The inclusion of measures to facilitate such 'carbon offsets' was in many quarters highly controversial, but in order to get a deal, it was included.

Despite achieving this concession, domestic politics in the USA prevented much progress being made there in any event. This was in part because of the effect of the so-called Byrd-Hagel Resolution of 1997, which came from an overwhelming Senate vote (95 to nought) against the ratification of any climate agreement which placed obligations on developed countries, such as the USA, but without comparable obligations on developing ones (such as China). During that period it also became clearer how previous global agreements on ozone depletion contrasted with what was needed on climate. With ozone, a few technological shifts were sufficient to fix the problem. By contrast, addressing climate change implied implementing deep change across multiple sectors from energy to farming, and structural shifts being made to economic ideas and social systems.

Against this backdrop, George W. Bush withdrew the USA from the Kyoto Protocol when he took office in 2000. This move away from international climate action was subsequently reversed when Barack Obama was inaugurated as president, but then repeated by Donald Trump when he decided to remove the USA from UN climate accords in 2017, a policy that was reversed again when Joe Biden won the 2020 presidential election, only to be overturned once more with the election of Donald Trump in 2024.

Despite the political challenges, Kyoto did make a difference. The EU and Japan committed to reducing greenhouse gas pollution by on average eight per cent (across a five-year period from 2008 to 2012) compared with 1990, while the USA and Canada committed to seven per cent. These commitments galvanized debate within countries and helped drive some level of modest change. But, as was the case after Rio, in the post-Kyoto world it was evident that much more was needed.

This realization began to make an impact on what politicians were prepared to sign up to. In the UK, Labour had a carbon reduction goal in its 1997 manifesto, with those policies strengthened at the 2001 and 2005 elections. Across the EU, too, different countries, led by Germany, had set what were for the time quite ambitious goals. Across the environmental movement we called for the policies in play to be strengthened. This in turn was backed by new scientific analysis. As time went by this only became more certain, and more worrying.

In 2007, the IPCC published its Fourth Assessment Report, which concluded that '[m]ost of the observed increase in global average temperatures since the mid-twentieth century is very likely due to the observed increase in anthropogenic greenhouse gas concentrations'.[7] This finding came at a time when climate change was no longer a distant theoretical risk, but increasingly a clear and present danger, already manifest in real-world events, including the impacts of massive storms, droughts and heatwaves.

In that same year, I joined a team of Friends of the Earth International campaigners at UN talks in Bali, Indonesia. Alongside action on fossil fuels, there was a renewed emphasis on action to halt deforestation. A new set of talks opened on reducing emissions from deforestation and forest degradation (REDD), to be concluded at a major summit in Copenhagen in 2009. I travelled to the Danish capital with Thom Yorke of the rock band Radiohead. We had worked together during Friends of the Earth's Big Ask campaign which had resulted in the 2008 Climate Change Act: the first national law in the world to demand science-based reductions over a period of decades. We'd decided to take the message to the global stage, encouraging other countries to make a similar step. But despite the clamour for action, the talks in Copenhagen collapsed. Trust between nations evaporated over disagreements about money and who should take what action by when, exacerbated by mistakes made in the conduct of the talks by the Danish government.

After the Copenhagen disaster, it took six years to get fully back on track. It wasn't until the successful COP21 hosted by France in 2015 that we finally achieved a new global step forward on climate. COP21 took place in Paris two decades after COP1 in Berlin, while I was working as an adviser to the Prince of Wales (now King Charles). He did everything he could to help make the Paris climate summit a success, hosting groups of companies and countries at meetings and seeking to create a common cause among them. All this set the scene for success, and his leadership was recognized when he was invited to make the opening address.

COP21 succeeded where others struggled for several reasons. In Copenhagen, world leaders only joined for the last few days, by which point the disagreements between national teams were too entrenched for the leaders to fix. By being there from the start in Paris, presidents and prime ministers were able to agree broad trajectories for their negotiators to work through for the two-week duration of the summit. There had also been a major diplomatic effort to seal as much consensus as possible in advance of the actual summit, including in a number of preparatory meetings led by France, in which public pronouncements among key countries were encouraged, including between China and the USA.

Another factor contributing to its success was the overall goal represented in the target of keeping global temperature increase to well below two degrees centigrade, and as close as possible to 1.5 degrees. It was agreed that this goal could be pursued via countries determining their own contributions. This contrasted with previous meetings, including Kyoto, where the whole world had needed to agree on all the individual national emissions reduction goals. Subsequent COPs worked on refining the Paris summit's conclusions and encouraging greater ambition in nationally determined emissions reductions.

The impetus at Paris and during the international meetings that followed came from a combination of ever more urgent warnings from the climate-change scientists and rising public demand for political action, strengthened by the visible changes taking place that people could actually see and experience, from more violent storms to heatwaves, and from droughts to epic forest fires.

Against this backdrop of visible climate change, the Sixth Assessment Report from the IPCC published in 2021 reached a series of stark conclusions, including that '[e]ach of the last four decades has been successively warmer than any decade that preceded it since 1850', that '[h]uman influence has warmed the climate at a rate that is unprecedented in at least the last 2000 years', and that the '[g]lobal mean sea level has risen faster since 1900 than over any preceding century in at least the last 3000 years'.[8]

In a series of future scenarios, the IPCC projected how different levels of action on emissions levels would lead to different outcomes. In the worst-case scenario, based on a plausible future in which we carry on with high emissions, an average global temperature increase of above five degrees is an outcome that is in the realm of 'very likely'. This would be literally catastrophic – for the economy, for civilization and for the integrity of the ecosystems that sustain life on Earth. A further report from the IPCC on the impacts of climate change published in 2022 set out the situation very clearly, stating that 'any further delay in concerted global action will miss a brief and rapidly closing window to secure a sustainable and liveable future for all'.[9]

NATURE AND DEVELOPMENT

As was the case with the UN's climate convention, the biodiversity agreement thrashed out at Rio in 1992 was also subject to further refinement. In 2002, ten years after the Convention on Biological Diversity was opened for signature, the countries that were members of it adopted a Strategic Plan to deliver on its goals. The plan ran to 2010, with the overall aim of halting the loss of biodiversity by that date. It didn't work, and so in 2010 the plan was refreshed, with a comprehensive new framework agreed that year in Nagoya, Japan. There were 20 targets, and the idea was that these would set the frame for action globally and within countries for the decade up to 2020.[10]

Many other goals and targets have been agreed internationally, including on a subject that is fundamental both to halting and reversing the loss of biodiversity and to effective action on climate change – namely, the state of the world's forests. This had been one of the really major issues on the agenda for decades, including when I went to lead the rainforest work at Friends of the Earth back in 1990, and it had attracted a lot of interest at the global level ever since, even if countries were unwilling to negotiate a UN agreement to create a coordinated international response. Nonetheless, later on, and via other discussions, targets were set to slow down and halt deforestation, including via non-binding agreements adopted in New York in 2014 to halve the rate of deforestation by 2020 and to halt it entirely by 2030.[11]

At the time of the Rio summit, and in the wake of *Our Common Future*, talks highlighted the fundamental linkages with matters of poverty and development, and general commitments on these subjects were adopted. It took until 2000, however, for these to be codified into a set of more specific goals. This was in the form of a package known as the Millennium Development Goals. The eight goals were set for a 15-year implementation period, and included in the package were plans to make progress on subjects ranging from halving extreme poverty to halting the spread of HIV-AIDS and ensuring all children received at least primary education. In 2015 these were replaced by a new set of Sustainable Development Goals (SDGs) with a target date of 2030.[12] These set out to achieve not only social progress, but also closely related environmental outcomes as part of the same plan.

This was a critically important step, considering how the discussions in Rio repeatedly affirmed fundamental connections between environment and development. The SDG package of commitments was encapsulated in a 17-goal plan backed by 169 more specific targets. The idea was to bring poverty, climate change, biodiversity, gender equality, food security, the role of business, education and much more into a single programme. It has been described as the closest thing humankind has to a strategic plan for the world, and it quickly achieved momentum, picked up not only by governments to shape their policies, but also by private-sector companies, investors, international agencies and charities.

To return to the question that I began this chapter by asking, it's obvious we are not lacking in ambitious plans agreed at the global level, but the world has struggled to achieve progress in what was set out in these plans. With many environmental targets agreed, and some of the most important set more than 30 years ago, it would be fair to assume that greenhouse gas emissions would be falling by now and the natural world entering a period of renewal, with both of those contributing to social progress. So is that what is happening?

Take climate change. Emissions have continued to rise year on year. In 1992, the year of the Rio Earth Summit, global carbon dioxide emissions stood at about 23 billion tonnes. By 2019, when the science had for decades said emissions should be falling fast, it reached 37 billion tonnes.[13] In fact, since 1990, when the IPCC published its first assessment report, the world has released about as much carbon dioxide as it did in the whole of human history up until that point.

Although this is undeniably a collective failure, some countries and regions did indeed meet the modest original goals. For example, the European Union (EU) met its original commitment to stabilize its greenhouse gas emissions at 1990 levels by the year 2000, and even cut them down a bit, by about three per cent. As time has gone by, however, the gap between what is needed and what is being done has become progressively wider.

In 2022, the Emissions Gap Report found that even in the wake of the largely successful COP26 meeting in Glasgow the previous year, the world was very far from achieving the Paris Agreement goal of limiting warming to below 2 degrees centigrade, and preferably to 1.5 degrees centigrade.[14] Assuming countries implement what they have said they will do (which is a heroic assumption considering past performance), this would still lead to a 2.8 degrees centigrade increase by the end of the century. If the promises made to go beyond existing policies were honoured, that would only reduce this to a 2.4–2.6 degrees centigrade temperature rise by then. This would lead to potentially disastrous social and economic consequences, not to mention huge environmental damage.

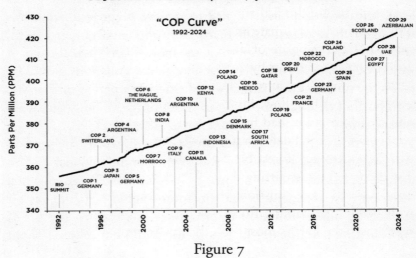

Figure 7

When it comes to the targets on biodiversity and the natural world, progress has been similarly weak. None of the 20 goals adopted in 2010 for the protection of life on Earth by 2020 were fully met at the global level, and only six of them were partially achieved by 2020, including increasing the proportion of land and oceans designated as protected areas.[15] Little to no progress was made on others, including the commitment to eliminate the hundreds of billions of dollars paid in subsidies that harm the natural world, such as those backing destructive industrial farming (although the UK has made some positive progress on this in the wake of its departure from the EU).

As a result of this failure to act, ecosystems have degraded at an unprecedented rate, driven by more land being cleared for agriculture, as well as climate change, pollution, and damaging practices such as fishing and logging that also have a direct destructive impact on wildlife. If we carry on as we are, the loss of biodiversity is anticipated to accelerate during the coming decades, and the continuing losses will of course hit an already depleted natural world. We have reached the point where only about a quarter of the habitat that originally existed on ice-free land is still functioning in a nearly natural way, and much of this is in protected areas or in those with low human populations.

On the subject of forests – the focus of the Friends of the Earth campaign before the Rio summit, and an issue that is so vital for outcomes for both climate and nature – progress has been not only weak, but the opposite of what is needed. Despite a target set in 2014 to halve forest loss by 2020 and to halt it completely by 2030, in 2019, with just a year to go before the 2020 deadline, a report was published revealing how, instead of forest loss being reduced, the rate of deforestation had increased, by a huge 43 per cent. Ninety per cent of the total deforestation took place in what are the world's two largest tropical forest blocks in the Amazon and the Congo.[16] The deforestation occurring there and elsewhere between 2014 and 2020 released more emissions than all sources across the entire EU, with nearly half coming from the destruction of primary tropical rainforests, the Earth's most ecologically diverse terrestrial systems.

When it comes to social progress, there's better news with the Millennium Development Goals, widely considered to be the most successful anti-poverty movement in history, credited with lifting more than one billion people out of extreme poverty, with increased access to drinking water and improved sanitation cited as two of the areas where significant progress was made.[17]

The more comprehensive SDGs, which included ecological as well as social aims, also achieved progress in certain areas. For example, in the least developed nations, the percentage of children attending school has gone up and there has been significant progress in reducing mortality among under-five-year-olds. Progress on environmental issues has been far more limited though, and in many instances it has actually gone in the wrong direction. These include increasing amounts of waste, the carbon emissions mentioned earlier, deterioration of coastal waters, overfishing, biodiversity loss and wildlife poaching and trafficking.[18]

BLOCKS TO PROGRESS

There's no question that progress is being held back by the influence of vested interests. The role of those organizations in slowing down action has continued, both within individual countries and also in global discussions. Indeed, at COP28 in Dubai in 2023, they were there in force; thousands of fossil fuel lobbyists seeking to influence the official delegates representing the member countries of the UN who were negotiating on the subject of phasing out fossil fuels. They outnumbered the representatives of the ten most climate-vulnerable countries and worked with vocal petro-states, including Saudi Arabia and Russia, to weaken resolve to phase out fossil fuels.

On top of this pushback because of economic interests, there has been a heavy slug of ideological rejection of climate action, often characterized in terms of 'freedom', 'growth' or affordability. Donald Trump's gleeful tearing up of national laws and withdrawal from international environmental treaties is a prominent part of his political offer. In a similar vein, the former president of Brazil

Jair Bolsonaro (who famously called himself the 'Tropical Trump') embarked on plans to reverse protections in the Amazon rainforests and promote development via land clearance. The political success of these men reveals how such ideas can still gain traction.

These anti-environmental narratives can generally be traced to fears (often more imagined than real) about the perceived threats to economic goals, restrictions to business opportunities and limits on personal freedoms. At times, it appears very much as if the rejection of environmental ambition is part of the same political project as that which seeks to protect the interests of the already wealthy. Money invested in coal mines, oil fields and forested land earmarked for conversion to commodity crops is at risk when environmental controls seek to limit all of that, and repeated well-funded political campaigns have been deployed to protect the status quo.

The ongoing flurry of anti-environmental sentiment expressed by prominent politicians in different parts of the world underlines the extent to which the science of climate change sometimes doesn't connect with the key decisions needed to avoid disaster. The political world exists in parallel with the scientific one of data and reality, and in the former choices continue to be made that take us away from the path recommended by a vast body of environmental science.

The fact that climate-change indifference and denial can sway voters confirms how vulnerable public backing for environmental ambition is to shifting political narratives. This in turn can be traced to, among other things, climate-change denial being widely touted in the media and social media, with such views repeatedly fed into the public discourse by prominent commentators, despite the absence of supporting evidence. The twisting of science through fake facts and outright lies is not immediately obvious to many hearing the messages, not least because of the complexity of the subject matter and the impenetrable nature of the jargon. Skilfully delivered populist rhetoric often seems to make more sense than the data carried in the pages of peer-reviewed scientific journals. Newspaper headlines like 'So much for global warming!' during a cold snap prey on popular misunderstandings about the difference between weather and climate.

This gap between popular opinion and proven fact exists in other areas than climate change – it is also linked to the demise of the natural world. Although denial of this problem has been less evident, the language and concepts used are sometimes rather technical and off-putting for non-specialists. For example, one vox pop survey undertaken among random members of the public to assess their understanding of the term 'biodiversity' revealed the most common answer was 'some kind of washing powder'.[19] Such is the chasm between most citizens and their appreciation of the state of the declining biosphere upon which we, as well as the rest of life on Earth, depend.

It is also widely assumed that the solutions to climate change (and by implication other environmental challenges) are technological, the idea being that we can carry on as we are in the expectation that something will be invented to solve all the problems. On the climate-agenda, nuclear fusion and more recently hydrogen technologies have been prominent in this space, as have (rather more realistically) electric vehicles and related battery technologies. Technology development is vital, especially for solutions that can be quickly scaled up, but if they are seen as fixes in the absence of deeper systemic change linked with economic and social transitions, then they are at best partial remedies and at worst dangerous distractions from the real issues at hand. It is also important to note that we already have a range of proven technologies and know-how that are not being used at the scale they could be, from wind turbines to recycling, and from regenerative farming to the construction of more energy-efficient buildings.

These and other barriers to progress have contributed to the fact that, despite the existence of a massive body of scientific knowledge and a vast political effort having been made over decades to embed solutions to environmental challenges, we are nowhere near where we need to be. One figure from history who might offer helpful wisdom is Dwight D. Eisenhower, commander of the D-Day landings and later the US president. 'If a problem cannot be solved, enlarge it,' he said. Despite the gargantuan nature of deeply challenging environmental pressures, it might help if we viewed them on an even broader scale.

For, on top of the pushback from vested interests, the ideological rejection of environmental science, the widespread public misunderstanding and the framing of the problem as technological, there is also another obstacle. It is a barrier to progress that has, in my estimation, been overlooked, although it was picked up in the review of the SDGs that reported underperformance in dealing with climate change and declining biodiversity. This additional obstacle is the failure to make serious inroads into tackling inequality.

The capture of a high proportion of wealth by an enriched minority diminishes the prospects of the majority, especially when, in order to maintain their position, those at the top divert resources away from sustainability programmes. So serious is the impact of the failure to deal with such issues that the authors of a 2019 review of the SDGs suggested that inequality threatened the entire sustainable development agenda.

The report uncovered how inequality can lead to greater social instability in the wake of shocks, pointing out how those at the bottom of society face increased risks arising from environmental degradation and biodiversity loss, and highlighting how this vulnerability increases the risk of migration and conflict. The relationship between inequality and the wider sustainable development agenda is complex, the report noted, but it is very real, and acknowledging it is fundamental to making progress. Delving into those linkages is where we go next, starting with a look at just how unequal we are.

3

Never Had It So Good?

Political turmoil is not unique to the twenty-first century. The late 1950s and early 1960s were also a time of volatility. Back then, policy disagreements led to the resignation of senior ministers in the UK government. The Profumo scandal, with its ramifications for national security, rocked the establishment. There was a looming battle with inflation. Debates on the best ways to sustain economic growth raged. Strained industrial relations caused political tensions. Foreign policy crises, including the Cold War, threatened global security like never before. It is interesting to note, then, that one of the most famous quotes from Harold Macmillan, from 1957, during his first term as prime minister, should be his claim that the British people had 'never had it so good'.

Macmillan was a 'One Nation' Conservative, working to a political philosophy that combined the maintenance of traditional values and institutions with progress for ordinary people. According to this worldview, the privileged and better-off should work towards the betterment of others, rather than seeking more advantages for their own social class. He spoke out against the high unemployment rate among his constituents in the town of Stockton-on-Tees, which he represented in Parliament. During his time as prime minister, average living standards rose, while numerous social reforms were advanced – for example, in relation to housing, working hours and child benefits.

It was in a speech at a Conservative Party rally in Bedford in July 1957 that Macmillan portrayed a situation in which, he claimed, 'You will see a state of prosperity such as we have never had in my lifetime – nor indeed in the history of this country.' He went on, 'Indeed let us be frank about it – most of our people have never had it so good.' He painted a picture of a country where the march of progress was helping ordinary people in ways that had not been seen before. One highly visible measure of change was the surge in private home ownership, which, during the 1950s, went up from 31 per cent to 44 per cent.

It was twelve years after the end of the Second World War, and Britain was indeed emerging as a different country compared with the one it had been when that conflict began. For one thing, its centuries-long imperial era was moving to its close. India won independence in 1947, and a cascade of other colonized nations were, during the 1950s and 1960s, going the same way. The post-war Labour government led by Clement Attlee initiated sweeping changes with the nationalization of strategic industries, including coal, railways, gas and electricity. Perhaps most famously of all, the Labour government that took office in 1945 instigated a modern, universal welfare state that provided free social security, healthcare and a range of social services.

In the USA, too, the post-war period saw rapidly improving conditions for many ordinary citizens. Car ownership exploded, as did the suburbs that depended on automobiles to connect the millions of new homeowners with their places of work and the services they needed. Consumer goods flooded from factories, some of which had previously been making tanks and guns; televisions, vacuum cleaners, refrigerators and washing machines were among the new must-haves. As was the case in the United Kingdom, Americans were more than ready to embrace the new opportunities for comfort and novelty following the years of shortage, tragedy and drudgery that came with war and its aftermath.

Unemployment was low and wages high. With more things to buy, and more money to buy them, the economy boomed. Between 1945 and 1960 the gross national product (GNP) of the United

States more than doubled, initiating what came to be known as a golden age for US capitalism. As in the UK, the US government had also instigated a series of social reforms, although not as extensive, generous or comprehensive, that helped create a sense of optimism for the future. This in turn was reflected in what subsequently came to be known as the 'baby boom', with more people having children and average family sizes increasing. About four million babies were born each year during the 1950s, and by the time the trend tailed off in 1964, almost 77 million 'baby boomers' had been added to the US population.

To supply this rapidly expanding demand driven by the rising population and expanding economic growth, which were also features of some of the emerging nations of the Global South, more resources and food were required. Factories that had been making the explosives that blasted the way to victory for the western allies switched to the manufacture of agricultural fertilizer. Some of those making poison gas and other essentials of modern warfare turned to producing pesticides. Both chemically based farming strategies were harnessed alongside new and bigger machinery, advanced plant-breeding techniques and land conversion to drive the 'Green Revolution', that was aimed at radically increasing food production. By dramatically ramping up output, the plan was not only to increase food security but also to slash food prices, heralding an era of falling costs. Changing tack from a battle against totalitarian regimes, an all-out conquest of nature was unleashed.

This was the first decade of the Great Acceleration. Soon the post-war explosion in consumption (at least in the developed industrial countries), economic growth, population and all the rest began to make its impact felt, both positively, in terms of improving social conditions, as well as negatively, as underappreciated environmental consequences started to make themselves felt. Rachel Carson published her seminal *Silent Spring* in 1962, but no one spoke then as they do now about global heating, a gathering global mass extinction of animals and plants, or resource depletion, although the deep foundations for these crises were being laid as the acceleration gathered pace.

Technology was the basis for many people's greatest hopes for the future. Mass production created a new range of goods available to more and more people, powering a consumer boom the likes of which the world had never seen, and which has continued and spread unabated ever since, despite the ups and downs in the rate of economic growth. But one important reason why Macmillan was able to make such a positive claim at a time of multiple political pressures and challenges was that during those post-war years this technology and consumer boom had been harnessed not only to drive economic growth, but also to reduce at least some social inequalities.

MEASURING EQUALITY

Equality and inequality are notoriously difficult concepts to pin down, or to reach definitive conclusions about. Individuals suffering from the effects of inequality experience a range of disadvantages, many of which interact with one another. Even in relation to income, there are many ways to measure inequality. Is it best to look at the income of households, or at income per capita? Is it most instructive to consider the differences between the richest and poorest within different societies, or should we first and foremost consider income differences between countries? Then there is the difference between wealth and income, and how best to account for the fact that some people with a seemingly low income own property and other assets that, if liquidated, would be worth many multiples of the annual average income.

All of these measures can of course be instructive, each telling us different things, and over the years various methods of putting numerical values upon them have been refined. One of the most widely used methods to quantify equality was put forward by the Italian sociologist Corrado Gini in a 1912 paper entitled *Variabilità e mutabilità* ('Variability and Mutability'). Gini's basic idea was to calculate a single number to describe the distribution of wealth or income across a society, with zero describing a perfectly equal society in which each household has an equal share, rising to one in a situation where a single household has it all and the others have nothing (it is also sometimes presented as a percentage ranging between 0 and 100). Countries with a lower Gini coefficient are thus more equal, and over time a society with a declining Gini coefficient is moving towards a more egalitarian situation.

Like all single numbers harnessed to describe complex situations, there are limitations to what the Gini coefficient can show. For example, being a relative measure, it does not reflect a country's overall wealth or income, only its distribution. For this reason, a very rich country and a very poor one can have the same Gini coefficient. It is also only as good as the data used to make the calculation, and in places with less reliable information it may be less accurate. As data collection methods vary and might change over the years, the situations between different countries might not be comparable. In some countries, data are collected on income, and in others on consumption, with the latter generally a better indicator of social well-being, especially in developing countries. This can limit comparisons between nations that are measuring different things. Despite this and other qualifications, the Gini coefficient is nonetheless a broadly helpful measure of trends and does enable at least general comparisons to be drawn.

As far as the UK is concerned, the changing Gini coefficient reveals that overall economic inequality was declining in Britain throughout most of the twentieth century following a peak in income disparity around 1910, just before the First World War. That peak was reached after more than a century of wealth concentration that had accompanied the Industrial Revolution. In 1800, the wealthiest 20 per cent of British citizens had a 65 per cent share of the income (a Gini coefficient of 0.60), with the wealth concentration reaching its greatest extreme between 1900 and 1910, at which point the top ten per cent of the property-owning classes controlled 94 per cent of the nation's wealth (with the top one per cent controlling 70 per cent). It took a brutal war to begin a process that started to level out this extreme situation, and over time this led to an era of wealth equalization on a scale never before seen.[1]

This process gradually gathered pace during the inter-war years up to the 1940s, and then accelerated, with the richest cohort of British society seeing, by 1990, their share of national wealth almost halved compared with 1910, while the 40 per cent or so of the population in the middle classes saw their share rise to 42 per cent. Even the lowest half of the population now shared ten per cent of Britain's wealth, while the top one per cent held 18 per cent, marking a massive transformation towards the equalization of society. The changing Gini coefficient marked the scale of the change, culminating in the low (i.e. most equal) point of 0.24 being reached in 1979.

The USA too enjoyed a period of greater equality during those post-war years, having followed a similar pattern during the twentieth century, with a high point in income inequality reached during the late 1920s, when almost half of all national income went to the top ten per cent of society. During the Second World War, the share held by the top ten per cent fell sharply to about 32 per cent and then hovered around that level until 1979, when, as was the case in the UK, it began to climb again. The late 1970s and early 1980s marked a breakpoint, when the post-war consensus ended and new economic ideas began to gain real traction in politics, leading to a focus on increasing the per capita GDP, with less emphasis on how wealth creation was distributed.

And so it was that from the 1980s onwards, in the UK and the USA, the trend of the earlier part of the twentieth century went into reverse. Since then, the tendency towards greater economic inequality has continued into the twenty-first century. Indicative of these longer-term changes, in the UK the Gini coefficient went up (i.e. towards increased inequality) from the then low point of 0.24 in 1979 to 0.34 in 1990 and 0.36 in 2009–10. Since then, it has remained fairly constant. However, by remaining at this level it has effectively increased inequality because while the increase in the overall size of the economy has averaged about two per cent per year, with a Gini coefficient of around 0.35 each household in the top ten per cent of incomes takes over seven times the proportion of the national income compared with households in the bottom

50 per cent, of which there are many more. Each year that the Gini coefficient remains at that level, the sevenfold gap increases by about two per cent, over time making the inequality gap 81 per cent bigger than it was in 1990.[2]

A similar pattern has emerged in the USA, where in 2008 a level of inequality was reached that was even higher than it was at its peak in 1928.[3] This is reflected in a rising Gini coefficient, which in 1979 stood at 0.35, by 1991 had risen to 0.38, and in 2006 had reached 0.41, where it has more or less remained. In 2015, the top one per cent of earners in the USA averaged 40 times more income than the average earner in the bottom 99 per cent. The disparities between the economic top and bottom of US society are starkly reflected in the fact that, even in the world's largest economy, an estimated 12.4 per cent now live below the poverty line.[4]

A complex array of social, political, policy and economic factors have led to this change of direction, towards a less equal distribution of wealth. There will be more on that later, but suffice to say for now that the shift was manifest at many levels, including in popular culture. The 1987 hit film *Wall Street*, starring Michael Douglas and directed by Oliver Stone, rather summed up the changing mood and came to be seen as an archetypal portrayal of the excesses of the 1980s. Douglas's character, Gordon Gekko, an unscrupulous corporate raider who deploys financial strategies to take control of companies and then strip their assets, utters the much-repeated line, 'Greed, for lack of a better word, is good.' The movie not only summed up the mood of the times, but Douglas and Stone reported how people told them for years afterwards how they'd sought careers on Wall Street through being inspired by the film.

The shifts that took place in the UK and the USA during the 1980s, including the rise of the financial sectors, were indeed significant, but the UK and the USA are far from being the world's most unequal countries. In the early part of the twenty-first century, that dubious accolade went to South Africa, which in 2014 (the last year for which good data were available, and a year before the SDGs were adopted) had a Gini coefficient of 0.63, which is close to the UK's level of inequality at its peak around 1910. In South Africa, the richest ten per cent then owned 71 per cent of wealth, while the poorest 60 per cent of the population, which included the more than half of the entire country that lived in poverty, held just seven per cent. Other African and South American countries are similarly unequal, with Namibia having a Gini coefficient of 0.59, Suriname 0.58, Zambia 0.57, Mozambique 0.54 and Brazil 0.53.[5]

At the other end of the scale, in 2018, Slovenia recorded the lowest Gini coefficient of any country, at 0.24 (the same as the UK in 1979), followed by the Czech Republic and Slovakia at 0.25, with Iceland 0.261 and Ukraine 0.266. In Western Europe, Belgium makes the top ten of the most equal countries, by the Gini measure at least, with a score of 0.272. Overall, it is nations in Europe, and Nordic, Central and Eastern European countries in particular, that dominate the bottom end of the league table of Gini coefficients globally, but even within these countries great disparities exist.[6]

DECLINING POVERTY AND RISING INEQUALITY

While there has been a tendency in many countries during recent decades towards an increase in income inequality, the difference between countries as a whole (although still great) has tended to decrease as economic growth has enabled what were once poor countries to seemingly catch up with historically more developed nations. This is where there is some good news, compared with the situation during the 1960s and 1970s, when the 'Third World' was still a distinct subset of the global community, demarcated by high levels of poverty. Since the 1980s, average per capita economic growth in developing countries has generally exceeded that of the more advanced economies, resulting in incomes across the world tending, overall, to converge.

Behind this declining global Gini coefficient has been a steady fall in the number of people living in absolute poverty, although the progress made on that critical indicator has been driven by a few countries where rapid industrialization has largely mimicked what occurred in England 200 years ago, with China making an especially significant contribution. That nation has for several decades generated and sustained a high level of economic growth, driven by industrialization powered by billions of tonnes of coal, enabling the mass migration of poor rural people to factory-based work in fast-expanding cities. This style of development has helped enable the poorest 50 per cent of the world population to secure a significant increase in income, mostly arising from high economic growth in Asia.

But even now, following decades of target-setting and development effort (and before the impact of the COVID-19 pandemic and the recent cost-of-living squeeze is taken into consideration), data collated by the United Nations reveal how 1.2 billion people in 111 developing countries live in what that organization calls 'acute multidimensional poverty'. This definition goes beyond simple income measures to also embrace access to education, healthcare, fuel, power, sanitation, water and good nutrition as measures of poverty.[7]

Those who have just moved out of extreme poverty remain highly vulnerable to shocks that can and do push them back into hardship, especially as half of the global population do not have any form of social protection. The experiences of the poorest four billion people today are sometimes comparable to those of Victorian factory workers, who could get by while fit and well, with available work and affordable prices, but who struggled to survive when sickness or economic shock struck. As far as the level defining extreme poverty in income terms is concerned, the World Bank now expresses the threshold in 2017 Purchasing Power Parity (PPP) prices and judges the typical poverty lines to be US$2.15, US$3.65 and US$6.85 income per capita per day, respectively, in low-income countries (such as Madagascar and Afghanistan), lower-middle-income countries (such as India and Indonesia) and upper-middle-income countries (which include the likes of China, Brazil and South Africa).

That the number of very poor people has over time gone down (notwithstanding recent reversals) is to be celebrated, but economic inequalities have nonetheless gone up, as the richest have become considerably richer. This is in part due to the benefits of economic development having been allocated unequally. For example, the poorest 50 per cent who saw an increase in income received between them only a 12 per cent share of the global gains, while the richest one per cent took a 27 per cent share of the wealth created. This concentration of wealth has become increasingly prevalent. During the 1980s, the richest one per cent of the world's population had 28 per cent of total wealth, but by 2017, they had 33 per cent, while the bottom 75 per cent had stagnated at around only ten per cent.[8] So, while overall poverty has gone down, inequality has gone up.

According to the global anti-poverty campaigning organization Oxfam, economic inequality is now 'out of control', its 2020 report finding that in 2019 the world's 2,153 billionaires controlled between them more wealth than the poorest 4.6 billion people (the latter representing more than half of the global population).[9] Oxfam juxtaposed these numbers with the World Bank's estimate that almost half of the world's population lives on less than US$5.50 a day. It pointed out that while spectacular wealth was being accumulated by a tiny minority, the rate of poverty reduction was going down. In fact, the rate of poverty reduction has halved since 2013, and that was before COVID-19 hit. More on that shortly.

As the rich have become considerably richer and the share of global wealth going to the very poor has at best stagnated, the experiences of the middle classes in many countries have also changed. During the post-war decades, at least in western societies, the feel-good factor of rising incomes and expectations shaped the mood of many people, as did the seemingly relentless march of progress and technology. That has since changed, for while technological advancement has continued, the rise in incomes for the middle classes in Western Europe and the USA has been sluggish. In the USA, for example, while the productivity of workers has doubled since the 1980s, almost all the gains have gone to executives, company owners and the investors who put in the capital to enable the companies to grow and create profit, while at the same time the incomes of the workers have stagnated.

A 2022 review of the SDGs identified several factors contributing to this trend, including the effects of automation, economic globalization (which has pushed down wages by pitching the greater profits that come from employing cheaper workers in one country against the situation in competitor countries), the declining influence of unions, and stagnant minimum wages for public employees.[10] A similar picture has emerged in the UK and other developed countries, with job creation concentrated at the high- and low-skills ends of the spectrum, marking a progressive hollowing out of the middle.

Having said this, looking at the global picture, the richest ten per cent of the global population includes not only the super-wealthy, but also many of those in the middle classes of Europe, North America and other developed countries, who, although they are in the middle in those countries, globally rank among the top ten. Currently that ten per cent take 52 per cent of global income, whereas the poorest half of the population has 8.5 per cent of it. On average, an individual from the top ten per cent of the global income distribution earns US$122,100 per year, whereas an individual from the poorest half makes US$3,920 per year.[11]

Under these circumstances, it is sometimes more instructive to look at the differences between the richest and poorest members of societies, rather than at the differences between the richest and poorest countries, to get a true picture of inequality. In order to understand this gap better, a new measure was proposed in 2013 by economists Alex Cobham and Andy Sumner, named after the Chilean economist, José Gabriel Palma, based on his 'Palma proposition'. Referred to as the Palma ratio, it was a response to one drawback of the Gini coefficient, which is that it tends to underestimate change because of the relatively stable (notwithstanding the observations above) income of the middle classes. To overcome this, the Palma ratio excludes the middle to look at the 50 per cent of the population who are in the highest and lowest income brackets – namely, the top ten per cent and the bottom 40 per cent.

The ratio is calculated by taking the share of Gross National Income (GNI) of the richest ten per cent of a population and dividing that by the share going to the poorest 40 per cent. A higher Palma ratio indicates a greater degree of inequality between these income groups, with a higher proportion of national income concentrated in the richest section of society. Although the Palma ratio conveys a different measure of inequality, compared with that which is expressed by the Gini coefficient, at the global level a similar spread of countries emerges at the top and bottom of the range, with South Africa top with a Palma ratio of 6.8, and, at the other end of the spectrum, the Slovak Republic on 0.71, Slovenia on 0.83, the Czech Republic on 0.84, Iceland on 0.87 and Norway on 0.9.[12]

So, through various measures, it is clear that there has been a marked increase in economic inequality in recent decades. Susan Smith, University of Cambridge honorary professor of social and economic geography and a Life Fellow at Girton College, has looked at this and its effect on society through forensic data analysis since the early 1980s.

'The big picture is that historically over time there's been a concentration of the share of income and wealth among the top earners and top wealth holders in the population of the world, but also of many individual countries,' she told me. She observed how inequality declined from a peak during the early twentieth century, dropping in the 1970s and then rising since about 1980. 'There's a kind of U-shaped curve [. . .] and what a lot of people don't realise is that today we're looking at disparities in income and wealth that are heading towards disparities that obtained when the *Titanic* set sail and in the days depicted in *Downton Abbey*.' While the accumulation of spectacular wealth by a minority is celebrated by some, Smith describes the trend towards extreme inequality as the 'opposite of progress'.

'We've taken a step back by a century,' she notes. 'A very interesting feature of that is how, when we started to think about global inequality in a much more nuanced way, say in the sixties, as academics we were very focused on differences between countries, because the development gap was enormous. It still is in many ways, but between, let's say, the 1980s and the 2020s, disparities between countries were eclipsed by the magnitude of inequality within societies. [. . .] In terms of overall income and wealth, where you lived in the world used to matter most; now extremes of income and wealth are found in all societies.'

Smith told me how absolute poverty is increasing in certain areas where you would least expect it: 'In the United States the infant mortality rate in some parts of major inner-city areas is larger than in some African countries.' She went on to highlight a conceptual trap that comes with looking at the reduction of absolute poverty in some places, 'which, although positive, takes our eye off the ball of the damage that the high levels of inequality within nations, as well as between nations, [are] causing. It's leading to people in parts of the world that you would imagine were quite affluent to be in absolute abject poverty.' The fact of rising inequality being a global trend is underlined by the fact that once famously equal societies are now on the same broad trajectory. 'Even though Sweden has traditionally been seen as more egalitarian than most, trends in economic inequality still follow a U-shaped curve. These are global patterns in a very interconnected set of political and economic processes.'

Simon Szreter, professor of history and public policy in the History Faculty at the University of Cambridge, has also devoted a lot of time to researching inequality. He is acutely aware of the differences between poverty and inequality and, like Susan Smith, believes it is important to be clear about the differences between them. 'If we're talking about the UK, it's very, very clear that inequality has become quite hideous by an extreme, even by our own historical standards,' he told me. 'I think one of the problems, and why it's tolerated, is that people don't actually have a clear understanding of the difference between absolute and relative inequality and poverty and there is a feeling amongst a lot of the population that we don't have absolute poverty anymore in this country.'

Another reason why inequality is not leading to more demand for change could be the fact that our present situation has arisen from trends lasting over four decades. It may have become normalized through being a gradual process that now spans a couple of generations. In any event, there are times, especially during crises, when the scale of divergence between rich and poor is laid bare, including when society-wide shocks cause stresses and strains that lead to those with the least being most vulnerable to the negative effects that follow, even in countries with social safety nets.

At such moments, not only are the least well-off hit hardest, but inequalities can be deepened. According to the World Bank's 2020 Poverty and Shared Prosperity report, the Gini coefficient increased by about 1.5 points in the five years following major epidemics, such as H1N1 (2009), Ebola (2014) and Zika (2015).[13] While the effects of the COVID-19 pandemic are still being calculated, it was predicted that an increase in the Gini coefficient of 1.2–1.9 percentage points would occur, signalling a significant increase in income inequality.

Like other major epidemics, the effects of COVID-19 intensified social differences at a number of levels, and among the 18 countries for which 2020 data were available, two thirds saw the number of people on low incomes increase during that year, with projections suggesting that inequality between countries rose by 1.2 per cent between 2017 and 2021. Income inequalities within countries are also believed to have increased, while the shock waves created by the virus have also been one reason for rising costs for consumers, hitting those on lowest incomes hardest, as inflation soared.

In a report called 'Inequality Kills', Oxfam estimated that the incomes of 99 per cent of humanity were diminished because of COVID-19, with those living on the lowest incomes hit hardest.[14] At the same time, the rich got considerably richer. The ten richest men in the world doubled their combined fortune in the two years following the start of the pandemic in early 2020, from US$700 billion to a total of US$1.5 trillion. The World Bank reinforced this picture with its estimate that hundreds of millions of people in the developing world were 'reversed back into poverty', with up to 115 million falling back into extreme poverty because of COVID-19.[15]

I asked Danny Sriskandarajah, the then CEO of Oxfam GB, about Oxfam's findings. 'I think the more I look at it, the more worried I'm getting,' he said. Based on Forbes and other data sources, he told me that in 2020 and 2021, 'of all of the new wealth created in the world, two thirds has gone to the top one per cent of the world's population'. That same Oxfam report, published to mark the opening of the World Economic Forum meeting in January 2022, found that a new billionaire was created every 26 hours from the start of the pandemic, with the collective wealth of the then 2,755 billionaires surging more than it had in the last 14 years put together, rising to a combined five trillion dollars, marking the biggest increase in billionaire wealth since Forbes began compiling such numbers back in 1987. To put these astronomical numbers into perspective, it can be instructive to convert millions, billions and trillions into time. Counting to a million seconds would take about eleven and a half days, to a billion, just short of 32 years, while counting up to a trillion would take nearly 32,000 years. This vanishingly small segment of humankind is accumulating a hugely disproportionate share of global wealth. Sriskandarajah said, 'The rich are not just getting richer, they are getting richer at a faster rate than we've seen since records began.'

So it is that a tiny minority, just one per cent, controls billions – and collectively trillions – while more than half of the world's population shares less wealth than that combined wealth between them. With this astonishing context in mind, it is clear that the world is struggling to make progress towards the equality targets set out in the SDGs, including the aim to achieve and sustain income growth among the bottom 40 per cent of the population at a higher rate than the national average. Just how far we are from getting on track to reach that goal is underlined by the World Bank's World Inequality Report of 2022, which revealed how between 1995 and 2021 the top one per cent captured 38 per cent of the global increment in wealth, while the bottom 50 per cent captured just two per cent between them.[16]

It is believed by most economists that some level of inequality is a price worth paying for overall progress and growth which, in the end, benefits everyone. What is often less appreciated is the body of data that reveals how inequality is correlated with worse social indicators. This tendency is apparent across all societies, no matter what their average income, with the wealth distribution within a society mattering very much indeed. Research set out by Kate Pickett and Richard Wilkinson in their important 2009 book *The Spirit Level* revealed how in fairer societies there was a higher level of trust, together with more social mobility, better educational attainment and longer lifespans, while in less equal societies there were worse negative indicators, including in infant mortality, obesity, mental illness, teenage births, murders and rates of imprisonment.[17] These findings led the authors to conclude that inequality is bad for everyone, including the better-off. It also costs a lot of money.

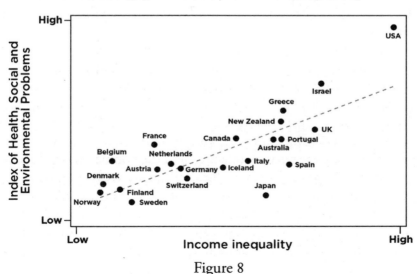

Figure 8

The Equality Trust has estimated the overall cost to the UK economy arising from inequality. It looked at four issues that we know are made worse by income and wealth disparity – namely, mental illness, physical sickness, imprisonment and murder. When compared with average rates across developed countries, the consequences of the UK's greater inequality cost the country about £106.2 billion per year.[18] And when the UK is compared with the five most equal developed countries, the cost rises to £128.4 billion. The trust points out that if other downsides of social inequality were added, the cost of our inequality would be much higher.

Perhaps this is why Macmillan's claim that the British people had never had it so good rang so true with many citizens back in the 1950s and 1960s. It certainly did for my parents, who'd grown up during the 1930s, and who'd experienced the consequences of the highly unequal society that had prevailed then. During the post-war years, by contrast, many industrial societies were on a path towards rising equality and all that went with it. Today, however, they are not, and neither are most developing countries, and as we shall see, this poses a major barrier to making progress towards other goals, including those needed to avoid environmental catastrophe.

LAYERS OF INEQUALITY

Wealth inequality is, of course, not the only dimension that determines the advantages that people can access. There are also factors linked with gender, race, sexual orientation, age, disability, religion, educational background and health. On top of these are also matters of inequality between generations, such as that between those baby boomers who enjoyed the benefits of ever-rising living standards, and their children and grandchildren, who are inheriting a depleted and less secure world, in part because of the consequences of previous generations' high-consumption lifestyles. Because of these differences, and depending on who they are, where they live and what their life experiences have been, people living in the same societies can have radically different prospects and outcomes.

We will turn to why and how these disparities are reflected in ecological challenges, but for now, let me share some brief thoughts on some of these differences and how they are manifested.

When it comes to the differences between men and women, one prominent area of debate has been in relation to income. When men and women do broadly similar jobs it has historically been the case that women have tended to be paid less. Despite widespread acceptance that this is unfair, and despite policies and laws designed to end such injustice, there continue to be cases that reveal the extent to which changing this remains work in progress. Some high-profile examples from the world of media have caught the headlines during recent years. One long-running story concerned the pay gap between male and female presenters working at the BBC.

In 2017, it was revealed in the BBC's annual report that two thirds of their top stars earning more than £150,000 were men, and that the top seven earners were all men too. Chris Evans made over £2 million in 2016–17, while Claudia Winkleman, the highest-earning woman at the BBC, earned under £500,000.[19,20] The BBC is, of course, not alone in this. Claire Foy and Matt Smith, who played the lead roles of the Queen and the Duke of Edinburgh in the Netflix blockbuster series *The Crown*, were arguably equally responsible for the series' success, but Foy was paid much less than Smith.[21] Whatever the reasons for this might have been, producers have evidently found that they can get away with paying female actors less than their male counterparts.

The same goes for publishing, an industry in which women authors are on average paid smaller advances than men and in which their published books are even sold for less than those by male authors. One study which looked at more than two million books revealed that titles by female authors were on average sold at just over half the price of those written by men.[22] These visible differences in the opinion-forming segments of society, in broadcasting and writing, are reflected more widely. According to the Office for National Statistics, in April 2022 the gender pay gap among full-time employees was 8.3 per cent, with the gap widening among those over 40 years of age and those in higher-paid groups.[23] Overall, according to the Fawcett Society, in 2022 British women took home on average £564 less per month than men (compared with £536 less in 2021).[24]

The injustice of men being paid more than female co-workers doing the same job is repeated in countries across the world, and across nearly every economic sector. In the early 2020s, the total share of global labour income going to women was about 35 per cent, which means that half the population shares just over a third of the income.[25] Although this reveals a massive disparity between men and women, the situation is better than it was in 1990, when women shared just 30 per cent of the total income.

Women have a higher share of total income in the former Eastern Bloc countries than anywhere else, with an average of 41 per cent. The Middle East and North Africa are at the other end of the scale, with an average of 15 per cent (the figure is under ten per cent in Saudi Arabia). One of the factors at play in creating the overall disparity is the low representation of women in top-paying jobs. This is beginning to change, however, with the percentage of women in the top ten per cent of the US wage distribution rising from 22 per cent in 1995 to 30 per cent in 2019. The proportion of Spanish women in the top ten per cent increased from 19 per cent in 1995 to almost 36 per cent in 2019, and Brazilian women increased their presence in the top ten per cent of wage earners from 24 per cent in 1996 to about 36 per cent in 2018.[26]

Why these countries have a higher proportion of women among the top earners is not entirely clear, although, paradoxically, it might have something to do with other aspects of inequality, in part possibly being linked with the affordability of private household help and childcare, which in turn reflects highly unequal wage distribution. Certainly in some countries, the UK included, the very high cost of childcare is a major issue in preventing women from returning to the workforce, and perhaps explains why the gender pay gap reaches its greatest level among 50-year-olds, when men have most of the best-paid jobs, with women having been held back earlier in their careers.

There is, of course, nothing new about the disparities we see today between men and women. Although the early hunter-gatherer societies had a high level of gender egalitarianism, when the transition was made to settled agriculture, gender inequality seems to have become structurally embedded. For example, evidence from Neolithic graves in Spain and China indicates that inequalities were beginning to appear as human groups shifted from hunting and gathering to farming, possibly due to the property and land ownership that came with that.[27,28] Genetic evidence gathered from Iron Age remains in Dorset, England, and was published in 2025. It suggests that society was organized around women, until it was replaced by the strongly patriarchal Roman system that came to dominate in the first century AD.[29]

Control of land has for millennia been linked with wealth and political power. In many societies land ownership was predominantly a male prerogative. Married British women with living husbands, for example, were not legally permitted to own land until the Married Women's Property Act of 1882.[30] Progress towards greater gender equality in politics was made during the twentieth century in many societies, although the situation varies widely across the world. Notwithstanding the twenty-first-century blots on the landscape of gender parity, there has since the end of the nineteenth century been positive progress towards less extreme gender inequality in how political power is wielded. But despite steady progress in many societies, in 2014 less than one quarter of the world's parliamentarians were women. In the UK, as of March 2024, there were 226 female MPs.[31] That's a little more than a third of the total of 650 seats, so although this is better than the global average, it is still some distance from equal representation with men. The figure changed for the better (to 263) later that year following a General Election, but whether this will be a permanent shift remains to be seen.

Racial discrimination, like gender inequality, can take a variety of forms, some explicit and others more subtle. The domination of one racial group over others took on global dimensions during the fifteenth century and gathered momentum in the centuries that followed, as Europe's colonial powers extended their tentacles of exploration and exploitation across Africa, Asia and the New World. European explorers went in search of land and resources, and found both in abundance. Much of the territory they found was of course already occupied, and had been for thousands of years, and these native societies were exploited as much as the land and natural resources.

Enslavement of African people took on industrial proportions, with millions forcibly transported across the Atlantic Ocean, and often worked to death in plantations in the New World. Vast fortunes were made by the new owners of land seized from the native people. In the Americas, the native peoples not only suffered enslavement, they were also wiped out by diseases brought by the Europeans. Indigenous Americans had no natural defences to the pathogens that had lived alongside people in the Old World for thousands of years, and mass deaths followed from influenza, measles, mumps, smallpox and other diseases.[32]

Some populations of native peoples were slashed by 90 per cent, and sometimes the pathogens travelled faster than the invaders. Following the first contact with Europeans on the Atlantic side of South America in 1498, it took less than three decades before smallpox was wiping out the Inca civilization thousands of miles away. Inca chronicles reveal that their leader, Huayna Capac, succumbed to a disease which was likely smallpox sometime between 1524 and 1528, preceding the arrival of Spanish ships on the Pacific side of South America in 1532 by some margin.

While the spread of disease was for the most part inadvertent (although there are alleged stories of blankets infected with smallpox being deliberately distributed to native people), the enslavement of millions by the expanding colonial powers was not. By 1800, when the USA conducted its second census, nearly a fifth of the total population was comprised of slaves. A far higher proportion of slave ownership prevailed in the southern states, where about one half of the population was owned by the other half. It was much the same in South America at that time, where it is believed the enslaved proportion of the population was then about 40 per cent.

Although slavery was outlawed in country after country, the persecution of Indigenous societies continued. For example, in southern Patagonia, both Argentina and Chile continued to occupy Indigenous lands and waters well into the twentieth century. During the late nineteenth century, sheep farming and gold prospecting expanded into the lands of the Selk'nam, Yaghan and Haush peoples, where European settlers, in concert with Argentina and Chile, engaged in the systematic extermination of the native people in a campaign that is known today as the Selk'nam genocide.[33] A bounty was literally put on the heads of Indigenous people by companies that would pay sheep farmers and militia members for the presentation of the head of a man, woman or child.[34]

On the far side of the Pacific Ocean, a massive genocide was also unleashed against the aboriginal Australians, who, having preceded the arrival of the British in 1788 by about 65,000 years, had evolved an entirely unique relationship with the land across that vast territory. With many living in nomadic hunter-gatherer societies, there were between 600 and 700 cultural-linguistic groups, comprising a total pre-colonial population estimated to have been between 700,000 to 1.5 million.

Despite this historic occupation of the land, the British invoked the notion of *terra nullius*, or land belonging to no one, as a legal pretext for seizure and settlement of territory without having to negotiate or pay for it. With that concept as a backdrop, by 1901 fewer than 100,000 aboriginal Australians remained.[35] This drastic decline was the result of campaigns of persecution, with brutal practices that included hunting people from horseback and scorched-earth tactics designed to trigger famine and disease. Mass round-ups and shootings were organized, so were mass poisoning and the removal of children.[36] They were forced into human zoos in Europe, where people back home could see for themselves examples of the 'savage' societies encountered at the frontiers of empire.[37] So traumatic was the experience that most Indigenous people died within a year of entering a human zoo, although there is a story of a Selk'nam boy who incredibly survived the ordeal and made it back to Patagonia.[38,39]

Such exhibits were undoubtedly a contributory factor in laying the foundations for the racism that would infect imperial cultures, based on the portrayal of captive people as less human than their exhibitors, as 'savages', 'inferior', 'other', and objectified. After all, zoos were about displaying animals, and putting people on display was bound to have a cultural impact on those filing past the enclosures.

It might seem as if such practices were from a long-lost era, but the last human zoo, portraying a Congolese village with live exhibits, persisted to within living memory, closing in Belgium in 1958.[40] Other aspects of explicit racism persisted until recently too, including in the policies and practices of the USA, South Africa and Australia, the latter only doing away with its 'Whites only' immigration policy in 1966.[41] And while positive changes have been seen in many countries, with laws and policies enshrining the equal treatment of people regardless of their race, in many societies there is still embedded discrimination.

With deep historical and cultural roots, racism remains stubbornly present even in countries which, like the UK, largely regard themselves as inclusive and multicultural. For example, Black people are seven times more likely to die after police restraint than white people.[42] Black girls are twice as likely to be permanently excluded from school as white girls and are placed in lower sets than warranted by their ability. When attempting to enter the workforce, ethnic-minority candidates have to send 60 per cent more job applications to receive as many calls back as white British people. Ethnic-minority graduates are significantly less likely to obtain employment six months after graduation than white graduates. They will also on average earn 20 to 25 per cent less per year when they are employed.[43]

Across all sectors in the UK, there is a higher percentage of people of colour in lower-paid positions, with the proportion drastically reducing higher up the career ladder. People from Black, Asian and ethnic-minority backgrounds suffer from a pay gap compared with their white counterparts. That gap varies in scale across communities and the country, and while it has improved during recent years, it is still there. When it comes to the leadership of major companies, as of March 2022, only six CEOs of the FTSE 100 most valuable companies were Black, Asian or minority ethnic, and none were female.[44,45] Access to finance has a racial dimension too. During the decade 2009 to 2019, only 0.24 per cent of venture funding went to Black founders of companies, and only 0.02 per cent went to female Black founders.[46]

In politics, despite some progress, racial minorities have historically been under-represented in Britain's elected assemblies. The same can be said for public bodies, and for the people who run major organs of public life, such as media editors and judges. The field of conservation and environmentalism, a sector that generally embraces inclusive approaches and puts effort into diversity policies, is the second least racially diverse sector in the UK (after farming).

The impact of COVID-19 hit Black, Asian and ethnic-minority people harder than it did white people. In large part this was down to the jobs that people of Black and South Asian ancestry tended to hold, compared with those more usually taken by white people. People of colour were more likely to be in roles that by and large did not shift to home working (such as van, bus and coach drivers, shopkeepers, nurses, care workers, cleaners and hospital porters and chefs), leaving them more vulnerable to infection and thus higher average mortality. Other socioeconomic and environmental conditions compounded the peril. They were more likely to live in poorer housing and in overcrowded conditions, meaning there was no opportunity to isolate within the home. They were also more likely to live in areas of higher air pollution and suffer from underlying health conditions, which also made them more vulnerable.[47,48]

In local authority areas with twice the average number of people from ethnic-minority groups, death rates were 25 per cent higher than the overall national average. It is worth adding that no evidence has been found linking increased risk to genetic or genetically related biological factors. And against this unequal backdrop, young ethic-minority people were 1.6 times more likely to be fined under COVID-19 laws than young white people.[49]

More visible evidence of simmering racism in the UK exploded onto TV screens and social media platforms during the summer of 2024, when, following years of demonization of immigrants and Muslim people by prominent public figures, violence erupted on the streets across the country, especially in poorer areas suffering higher levels of social deprivation.

Inequalities in wealth and influence are tolerated partly because of the extent to which we believe that success ultimately comes down to individual effort, talent and ability. There is no doubt that this belief is to a large degree correct, although with some serious qualification in relation to the different starting points that people experience. One major source of difference is linked with education.

Good schools make a dramatic difference, and it is therefore no wonder that many of those with the means to do so pay huge fees for their children to reap the benefits. With smaller class sizes, excellent facilities, confidence-building cultures and well-to-do pupils, private schools have nurtured amazing talent that in turn has enriched society. The trouble is that only a tiny minority of people get to enjoy that advantage, in turn leading to massive inequality of opportunity experienced by young people from different backgrounds.

One UK study by the Sutton Trust and the Social Mobility Commission analysed the educational background of 5,000 people in top jobs across British society and found that they were five times more likely to have been privately educated than the average. Just seven per cent of the British population attends private schools. When it comes to top universities, only one per cent of those reading degree-level courses graduate from Oxford and Cambridge, with those elite institutions having a far higher proportion of privately educated students than the national average.[50]

An elite education pays dividends for those looking to assure privilege for their children. Nearly two thirds of senior judges (65 per cent) attended private schools, as did more than half of our top civil servants (59 per cent) and diplomats (52 per cent), as well as nearly half of our senior military officers (49 per cent), chairs of public bodies (45 per cent) and newspaper columnists (44 per cent). These are influential positions in society in terms of law, government policy, regulation and how the media shapes opinion and politics. The fact that such a narrow segment of people with such an elite educational background play such a dominant role in Britain's national fabric causes inequalities to be systematically embedded.

Those attending fee-paying schools not only access better teaching, they also have the opportunity to establish influential lifetime networks. Private schools concentrate pupils from similar backgrounds, enabling them to make the connections that lead to elite jobs in later life. Indeed, the former pupils of just nine of Britain's leading fee-paying schools are 94 times more likely to reach elite positions than those from other schools, and much of the edge they enjoy remains even when they don't go to university.

Being able to afford expensive school fees is of course linked with wealth and income, thereby creating a positive-reinforcement loop, where young people from wealthier families are put together at an early age, creating the conditions for a convergence in culture and worldview among future elites. From what I have seen, those elites are more likely to make choices that protect the system which created them, than to seek changes that diminish their influence and privilege.

It is worth adding that there has been a gender dimension here too, with Britain's most famous public schools, Eton and Harrow, still taking only boys in 2024. It was only in 1948 that the University of

Cambridge permitted women to obtain a degree, with Magdelene College holding out for decades after that as a male-only bastion, only finally accepting female students in 1988. Even then this basic step towards equality was regarded as a moment of great sadness, with its flag flown at half-mast.

THE GENERATIONAL DIMENSION

Irrespective of which school young people attend, anyone with many years of life ahead of them has good reason to question the fairness of decisions being taken today. I was born at the end of the baby-boom years and now have four young granddaughters. I sometimes wonder what kind of world they will see in the 2080s, when they will be the age I am now. As Johan Rockström pointed out, the Anthropocene epoch began during those same post-war years as the baby-boom era. Indeed, the great acceleration in consumption that took off at the same time as the post-war economic boom, causing global warming, deforestation and pollution, was in part down to the lifestyles of the baby boomers. And the boomer generation is still ramping up its legacy of ecological impact, in ways that are set to hit their descendants very hard.

Across Europe and North America, many boomers came of age in a time of increasing affluence, with widespread government support for housing, health and education, and growing up with a genuine expectation that the world would improve over time. The ecological crisis was by and large not visible to them then, and unending progress was assumed to lie ahead. They were told they'd never had it so good, and in some ways that was correct.

Contrast the boomers' experiences with those of youngsters today, with stagnating wages, unaffordable house prices, little prospect of a large pension in retirement (or of retirement at all), cripplingly expensive higher education and a world suffering from the visible effects of environmental instability. For some young people, the pessimism that has replaced the optimism of their parents' and grandparents' generations is leading to the opposite of a baby boom, with many choosing not to have children at all.[51]

Part of the reason for that is the hopelessness they see in the prospects for halting and reversing ecological degradation, which, despite all the warnings that have been in place for at least the last three decades, has remained largely unchecked.

Such are the many layers of difference, inequality and disadvantage that more or less shape societies across the world. Economic, gender, race and generational fault lines quite rightly attract attention in social policy, although it is rare that they are explicitly presented as being important for shaping environmental outcomes. However, the fact is that all these manifestations of inequality are linked with global heating, diminishing biological diversity, resource depletion and pollution. This is evident not least in how environmental degradation has disproportionately serious impacts on the least well-off and the most disadvantaged.

4

Affluence and Effluents

Carbon dioxide emissions get a lot of attention, but the industrial powerhouses that have driven our economic system for over two centuries have released into the environment much more than that. In the 1980s, by the side of the river Tees, there was a cluster of massive steelworks, chemical plants and shipping industries. Such was the scale of this vast complex of factories – and so impactful were the chimneys, plumes of steam, jets of flame, expanses of pipes, lights and sirens, and columns of pollution – that Ridley Scott, who lived in Teesside during his youth, was inspired to harness the spectacle as the backdrop to the opening of his dystopian 1982 film *Blade Runner*, depicting a future world afflicted by ecological collapse.

The vast Wilton chemical works that inspired Scott became the focal point for a Friends of the Earth campaign 15 years after his film hit the big screen. By then, I was helping to move along various pioneering projects that sought to link social inequalities with ecological challenges. One was a platform that set out to reveal the location of pollution sources. This work arose from the vision of my predecessor at Friends of the Earth, former Director of Campaigns Andrew Lees. He foresaw the rise of technologies that are now mainstream long before most appreciated their potential. He put the capability into the organization that enabled us to combine data management with GPS and the internet, with one early result being that pollution platform. We called it Factory Watch.

Factory Watch showed what pollution was being released and who was releasing it, but of course it was only as good as the data we had. During the early 1990s, that was limited. In those days, Her Majesty's Inspectorate of Pollution (HMIP) was the official regulator that gathered this information. It took a great deal of persistence from Friends of the Earth to obtain the data we needed, and to start with we drew a blank. We were spurred on, though, by our colleagues in the USA. In July 1992, our researcher Mary Taylor had teamed up with Friends of the Earth counterparts in Washington, DC to publish chemical industry data provided by companies operating there. Some of those businesses were also present in the UK, but while they disclosed information in the USA, they refused to hand over comparable information on the other side of the Atlantic, which puzzled us.

In the USA it soon became apparent that more data availability was leading to companies polluting less, helping us make the case for more official access to information in the UK. While we campaigned for more openness in the UK, a further major step was taken in the USA in the form of an executive order issued by President Clinton in August 1993 which instructed federal agencies to report any toxic releases into the environment and ordered federal bodies to reduce pollution as much as possible. We believed a similarly powerful breakthrough could be achieved in the UK.

Describing the situation in the UK during the early 1990s, Taylor recalled, 'HMIP had consulted on a Chemical Release Inventory of industrial emissions in 1992. We had criticized the plans because the data was linked to authorizations to release pollution which itself was inconsistent between plants even for the same industrial processes. So one plant might have to report a particular chemical, but another would not. The authorizations were also incomplete in any case because a new regulatory regime was being brought in bit by bit. It was laughable really. They totally missed the point about public access to information and accountability of polluters. They'd produced a five-hundred-page document on paper that cost thirty pounds. They had aggregated emissions by district council and failed to name a single factory. Some of the data only covered part years because of how new regulations were being phased in.'

At least this information revealed that HMIP was collecting data, and although it was presenting it in a partial manner and only on paper, if Friends of the Earth could get hold of it and present it in a more user-friendly format, its impact might be considerably enhanced. HMIP was resistant to releasing the data we wanted, but with the benefit of new laws, including a European directive on access to information (which we'd campaigned in support of), and through endless nagging of HMIP officials by Taylor, in the end we got the data.

Factory Watch was born. It was an online resource that would enable anyone in Britain to find their home on a map which used a scatter of red dots to display the industrial sites nearby. By clicking on a dot, they could find out the name of the industrial plant and details about the toxic releases emitted into the air, water and land. We provided full information about the pollution and health risks posed by 440 chemicals coming from more than 1,500 industrial plants and set out in plain language the risks from carcinogens, reproductive toxins and chemicals that could trigger asthma. As well as showing how new technologies could be harnessed to raise awareness about pollution, it was also a major step in opening up previously hidden environmental information.

Back in 1995, when Factory Watch was first launched, government action to share information about the actual state of the environment where people lived was not keeping pace with rising public awareness. Taylor told me how the release of Factory Watch impacted on the regulators. 'It really shocked HMIP. They went into crisis management mode, a situation not helped by the fact of them at the time only having one computer with internet access. Eventually they had to agree that publication of such data was an important step towards pollution management.'

Factory Watch marked the moment when for the first time ever people in Britain were able to easily find out which major sources of pollution were most likely to affect them and who was responsible for releasing it. We were able not only to generate simple language and maps, but to create league tables of polluting industries and chemicals locally, regionally and nationally. Factory Watch was also the basis for further analysis that Friends of the Earth undertook to reveal the extent to which pollution was linked to social inequalities. When we compared the relative distribution of pollution and poverty, we were shocked to discover that 82 per cent of the cancer-causing pollution released into the environment across Britain was concentrated in the poorest 20 per cent of locations. The poorer you were, the more likely you were to live in an area where the air that you and your children breathed contained toxic emissions.

We pondered how such a situation could have occurred. How had the affluent managed to so successfully isolate themselves from effluent? Was it because companies located their factories in places where people were already poor and thus had weaker voices in politics, and where the arrival of any kind of jobs, including in polluting industries, would be welcomed as a remedy to unemployment and poverty? Or was it because the better-off people moved away to escape the pollution? Other factors linked with race and housing policy might also be blended in too, as well as the momentum of history, which was certainly the case in Teesside, where heavy industry was not only regarded as a tradition, but was for some, such as those in the communities making steel (and previously ships), also a source of huge pride.

Whatever the combination of factors, the injustice was real, with those living in the lowest economic segments exposed to the highest pollution loads. The findings we'd unearthed mirrored more detailed work on 'environmental justice' that had been done in the USA. Researchers from academia, campaign groups and the US Environmental Protection Agency had found strong links not only between income and pollution, but also with ethnic minorities; moreover, a causal link was found in the siting of such industries in poor areas dominated by Black communities.[1,2,3] Researchers found that in such locations, industrial owners benefited not only from less opposition to the building of their plants, but also from weaker enforcement of environmental standards.[4] The effects of pollution in relation to ethnic groups were greater in the USA due to the tendency for housing policy to be based historically on the segregation of different racial groups.[5,6] Poor Black neighbourhoods were created and became the places where many of the most polluting industries were located.

Having travelled in the USA and seen first-hand some of the industrial complexes that dwarf those in Teesside, including the vast sprawl at Baton Rouge in Louisiana, I was struck by the predominance of Black people living next to the refineries and chemical complexes. Various reports going back to the 1980s revealed connections between race and pollution, including a 1987 study by the Commission for Racial Justice called 'Toxic wastes and race in the United States'.[7] This report drew on different data sets to paint a picture of how exposure to dangerous pollution was skewed towards the non-white community. It found that the neighbourhoods with the greatest number of toxic waste sites next to them had a high proportion of Black and other minority groups. Although socioeconomic status appeared to play an important role in determining the location of commercial hazardous waste facilities, race proved to be more significant still. This remained true after the study controlled for urbanization and regional differences. 'The possibility that these patterns resulted by chance is virtually impossible,' it concluded.

BUCKET BRIGADE

We took inspiration from American campaigner Denny Larson, who'd come up with a community-organizing approach that he'd called the 'bucket brigade'. This involved the use of a basic DIY testing kit that anyone could use in the wake of a chemical release. The idea was that, following a siren warning of a pollution incident, brigade samplers would run outside their homes, take an air sample in a bucket device and seal it so that it could be analysed and thus prove which pollutants had been released into the community. In the absence of proper monitoring and disclosure of what was being discharged and where, it was a practical way to gain some data and thereby power and influence with the companies and regulatory agencies. Denny came to Teesside to advise us and the community.[8] Another campaigner to visit Teesside from outside the UK was Desmond D'Sa from Durban in South Africa, who told us that the scenes he saw in Teesside were akin to what he knew from home, which he'd assumed didn't exist in developed countries.

Drawing on these kinds of experiences and perspectives, Factory Watch directly challenged the sense carried in some corners of the media and political thinking that environmental concerns were essentially middle-class preoccupations. According to this viewpoint, the environment was something that only the better-off worried about, whereas the interests of the disadvantaged would be better served through job creation, economic growth and investment in industry, even if it did generate pollution. We found though that the communities in whose name such policies were being advocated had little knowledge about what pollution was being released, nor much voice with which to express a view about what, if anything, should be done about it. But in helping to rectify this situation, we did confront major limitations in our own organization.

Friends of the Earth had an excellent network of about 200 volunteer local campaign groups scattered across the country, but we found that in the most deprived and polluted locations we lacked a local presence. This included Teesside, where, despite the pollution being serious, there was no organized local campaign working to address it, even when information about the extent of toxic releases was available. Why people lacked a voice there was not immediately obvious, but the pattern was repeated in communities across the country, whereby the more deprived the people were, the less able they were to respond to environmental pollution. I saw that at least part of the reason for this was down to how such communities, compared with better-off ones, lacked money, education and confidence, which meant that they were less likely to have a culture of joining groups like Friends of the Earth. Whatever the reasons, we realized that in many of the places with the greatest need we had the least ability to act, leading us to conceive of a new initiative to help those communities to respond.

American pioneers who'd linked the predicament of people living in polluted and degraded environments to matters of poverty and race didn't parachute in outside experts to make the case for an environmental clean-up. Instead, they empowered the communities living in the shadow of the industries to speak for themselves. We decided to do the same thing, and while we worked with very slender resources, we were nonetheless able to hire local community organizer Carole Zagrovic to set up and lead a new local social network called Impact. She lived in Teesside and got stuck into community work, organizing meetings, sharing information and seeking to engage people in pressing for better performance from major corporations that included DuPont, Ineos Chlor, Huntsman and BASF.

Carole had a different role to many of our staff, who were experts in technical, legal and policy matters. She worked directly with people on the ground, facilitating their voice, negotiating positions and enabling them to speak with regulators and industry. But with just one employed person and a community still finding its voice, it really did seem like an unequal struggle, pitting our tiny resources against the might of multinational corporations that had invested billions of pounds into Teesside. These companies had a lot of power and influence in local politics and among regulatory agencies. The challenge seemed even more one-sided considering some of the initial reactions we'd received from the community. For while we could see the issues raised by the pollution, many of the people who lived in the Teesside towns of Middlesbrough and Redcar did not, or at least didn't see pollution as the main problem they faced.

Elaine Gilligan led the development of our environmental justice activities, working closely with Carole in Teesside. She recalled some of the initial reactions from locals: 'You would literally have flares of pollution and sirens going off and an awful lot of industrial emissions, but when we spoke to locals about how concerned they were about that, they said they were more worried about the ice cream van dealing drugs. That was a big wake-up call for us in Teesside. They weren't scared of pollution. A lot of the local people who did have jobs and could afford to move out did so and then commuted in. The whole area became mostly unemployed people with no social facilities and lots of drug and crime issues. If you could afford to move out, you went, and so the area finished up with a high level of social dereliction.'

Against that backdrop of deprivation, environmental campaigners arriving at Teesside received a lukewarm welcome. It was not so much hostility, more perceived irrelevance. We faced criticism for our apparent lack of importance in relation to local issues and for our historical neglect of the area's pollution problems. Despite these initial reactions, however, we made progress and gained some credibility by exposing ourselves to local residents' doubts. As campaigns director, I was ultimately responsible for this work, and, being very aware of the novel nature of the activities we'd put into motion, I travelled to Teesside on several occasions to learn more about what we could do to ensure a positive impact.

By then, I'd been involved with environmental questions for quite a few years, but more as a scientist, policy advocate and campaigner, rather than working with directly affected communities. But although I was a professional environmentalist based in London, I did have some personal experiences with pollution. In 1984, my father died at the age of 61 from lung cancer, probably caused by exposure to asbestos at the Cowley car factory where he'd worked on the production line for many years. One of his best friends, Dennis, who helped foster my early interest in nature by taking me fishing on Sundays, died of the same mesothelioma a few years later. My uncle Tom, who'd lived a very healthy and active life, passed away from the same thing in 2009. They had all worked in the same car-manufacturing plant, and asbestos was linked not only to their deaths but also to those of other workers employed there. These experiences had left me with some vivid and disturbing images as to what toxic pollution could do to people.

We patiently plugged away on trying to engage the local communities, and as time went on, we did build trust. By enabling affected communities to develop a more informed and powerful voice for themselves, and as a result to demand stronger action from regulators and industries that had previously not acted with the vigour that might have been expected, pressure was now being applied in the right places, and to positive effect. Some of those working for the official pollution regulator, which was by then the new Environment Agency, acknowledged that the elevated and informed voice of the community forced the pace of action they took to deliver a cleaner environment. More widely, the very fact of shining a light on polluting emissions that had previously been largely invisible and poorly understood, even by those most affected by it, led to improvements.

A few years after we first launched Factory Watch, the release of the cancer-causing emissions that we'd highlighted dropped across England and Wales by nearly half (48 per cent), between 1998 and 2002. There were several reasons for this, and public exposure was certainly a part of it, with local and regional media coverage really putting pressure on polluting industries to clean up, especially when local people backed the demands. Eventually, the release of pollution data by the official regulator took a step forward, meaning that we didn't need to do that work any longer. In 2003, we closed the Factory Watch website, its objectives of releasing pollution data and raising awareness of environmental injustices having been largely achieved.

RACE TO THE BOTTOM

Alongside action from the companies thrust into the spotlight, there was another reason why pollution levels were falling in the UK: the process of globalization. The push for more economic growth driven by trade and investment through liberalizing the global economy was making it easier for many companies to reap cost-saving benefits by relocating to places where doing business was cheaper, because environmental laws were less demanding and wages lower. Indeed, many countries deliberately sought to keep environmental and employment regulations to a minimum to encourage international investment. These were among the drivers behind the rapid industrialization of major economies such as China, allowing them to take a bigger share of sectors previously located in the countries that had industrialized first.

The migration of chemicals, steel, shipbuilding, oil refineries and some manufacturing which began in the 1980s led to rapid decline two decades later across what were once heavily industrialized parts of Europe and North America. The rise of the so-called Rust Belt in the eastern states of the USA and the abandoned chemical and steel works across the UK were the visible consequences. For many communities, including Teesside, the pollution and poverty of the past was replaced by even worse poverty and unemployment in the present, not only exacerbating material wants, but also sapping trust, hope and pride.

The pollution had not disappeared, of course, it had just moved from poor communities in rich countries to poor communities in developing countries. We were very aware of this backdrop to the changing environmental situation, and when I was elected to the board of Friends of the Earth International in 2001, we saw new opportunities to reflect this shifting context in our work. The global federation of national Friends of the Earth organizations was by then scattered through more than 70 diverse countries, including El Salvador, Scotland, the USA, Indonesia, Nigeria, Norway, Sri Lanka and Colombia. Taking on the role of vice chair, I became involved with developing strategy. One key theme concerned the linkages between social justice and environmental degradation, evident in the situation faced by communities across the world whose lives were blighted by exposure to health-threatening pollution. What was happening was because of economic globalization, so we figured it was logical to globalize the idea of environmental justice. This led to a raft of campaigns, including one that would unite communities across the world living next to the operations of a single major company: Shell.

Shell was well known to us as a vast fossil-energy organization and one of the most massive corporations on Earth, with an annual turnover larger than that of many countries. It was based in the

UK and the Netherlands, and we had a strong Friends of the Earth presence not only in those countries, but also in many others where national partner organizations were campaigning on the ground to highlight the Anglo-Dutch giant's impacts. These included Nigeria, the USA, Caribbean countries, South Africa and Bangladesh. We wanted to unite them in a single international initiative to press for pollution reduction across the board, no matter where the fossil-energy giant operated. This became the Shell fenceline communities' campaign, bringing together people who literally lived on the other side of the fence to refineries, drilling operations, gas flaring and the rest of the industrial paraphernalia needed to feed the world's growing demand for gas, petrol, diesel and chemical feedstocks.

We brought representatives of these community organizations to London to attend Shell's annual shareholders' meetings, where questions were raised in front of their investors about the effects of Shell's operations on the people who lived next to its refineries and oil-production facilities. Brilliant campaigners delivered a united call for the company to do better. Among them was Oronto Douglas from Nigeria, where Shell's activities had been linked with multiple impacts on local communities, including the killing of Ken Saro-Wiwa, who had protested against the pollution caused by Shell's activities in the Niger Delta and paid the ultimate price when he was executed by the Nigerian government.

Desmond D'Sa, who'd joined us on visits to Teesside, came from South Africa, where the Shell-owned SAPREF refinery regularly released health-threatening pollution over the local community. Hope Esquillo Tura was a campaigner from a group seeking to remove a massive Shell oil and gas depot from next to the residential neighbourhood of Pandacan in the city of Manila in the Philippines. Margie Richard, an activist working to clean up the Shell chemical plant at Norco in Louisiana, was there too. Margie and her neighbours believed high rates of cancer, birth abnormalities and serious ailments, including asthma, were caused by pollution from Shell's operations near their homes. Hilton Kelley, from Port Arthur in Texas, with his broad Stetson hat, led a local campaign group committed to cleaning up the Shell refinery located next to his community. His group had organized health surveys conducted by the University of Texas Medical Branch at Galveston, which revealed that 80 per cent of surveyed residents in neighbourhoods near the refinery had heart conditions and respiratory problems, compared with 30 per cent of people in non-refinery areas.

Craig Bennett, a gifted campaigner who would later become Friends of the Earth's CEO, organized our efforts to bring the fenceline communities together. He worked with community leaders to combine their individual voices into a collective call for change. This work was deemed strategically important by Friends of the Earth because of the increasing tendency back then for global companies to signal their commitment to voluntary 'corporate social responsibility' (CSR). Shell was one of the most significant corporations to do so, and the concept was seen by many campaigners as providing a way for major companies to pre-empt and avoid binding laws, rules or regulations on social and environmental matters. As a leader in CSR, Shell also became a leader in the production of annual CSR reports to complement its annual financial report and accounts. The criticism from many, including the communities living alongside Shell's operations, was that these beautifully produced reports failed to mention some of the most significant impacts that the company was having on local people and the environment, and when they did, it was superficial and did not give a true account of what needed to be done to improve performance. At the time, and because of the partial picture they presented, these documents were seen as the epitome of 'greenwashing'.

This is why, during the early 2000s, Friends of the Earth produced a 'shadow' CSR report on Shell's activities for three years running, in which we told the real story from the perspective of the fenceline communities.[9] We distributed these reports to the media and Shell's shareholders, including at the company's annual general meetings. Our reports stressed the need for the company to take a view that widened its purpose from the pursuit of profit for its shareholders and to look to promoting the interests of the people whose lives its businesses impacted.

I asked Bennett about the Shell fenceline communities' campaign and what it set out to achieve. 'Obviously it was welcome when companies took genuine steps to improve their social and environmental performance,' he told me, 'but in many instances these CSR programmes were largely devoid of involvement from directly affected communities, even when they lived next to the polluting facilities. The purpose of CSR and the glossy reports seemed to be as much about countering calls for binding rules and regulations on social and environmental matters as it was to change anything on the ground.'

He continued, 'By producing our own shadow Shell CSR reports, we were able to tell what we believed to be a much more accurate story about the impact the company was having on the lives of its fenceline communities, and by bringing members of these communities to London to participate in the Shell annual general meeting, we were able to bring first-hand knowledge to counter the sometimes bland reassurances made by corporate executives.'

It worked. Bennett recalled the memorable intervention of Hope Esquillo Tura, the community campaigner from the Philippines, who asked Sir Philip Watts, the then chairman of Shell, when he was going to remove the company's massive oil and gas depot from the residential neighbourhood of Pandacan in Manila, as had been required by the local city authorities. 'Rather than offer a timeline for the removal of the facility,' Bennett recounted, 'Sir Philip tried to reassure Hope that the company was about to install what he called a "vegetative buffer zone" around the facility. Hope told the meeting that she had seen Shell's plans for a buffer zone, and it was to be just two metres wide. At that moment over a thousand shareholders at the Shell AGM burst into laughter, as the difference between the company's CSR gloss and the reality on the ground became plain for all to see.'

Questions of pollution justice underline the connection between social inequalities and the industrial pollution that comes from the processes that supply modern consumer economies. These injustices also manifest in the impacts arising from the use of the products produced by those industrial facilities, including petrol and diesel combusted in the engines of the billions of vehicles that run up and down the world's millions of miles of roads.

A BREATH OF FRESH AIR

Outdoor air pollution, much of which is caused by vehicles, is associated with damaging effects on lung development in children, heart disease, stroke, cancer, exacerbation of asthma and increased mortality. It arises not only from engine exhausts but also from particles produced by brake and tyre wear and is added to by agricultural pollution, including ammonia from livestock (which also causes major damage to sensitive ecosystems), and domestic, moorland and forest fires.[10] Some of it travels, too, crossing international boundaries. Globally it is estimated by the World Health Organization that outdoor air pollution annually causes the premature deaths of a great many people – in 2019, about 4.2 million of them (which is about 1.2 million more than died from COVID-19 in 2020).[11]

Around 89 per cent of these premature deaths occurred in low- and middle-income countries, with relatively fewer in the better-off nations. Differences in the impacts of air pollution can be seen not only between rich and poor countries; within individual societies, the less well-off suffer from a higher level of exposure to health-damaging pollutants. Across much of the world, driving along busy main roads involves passing by poorer neighbourhoods and blighted communities. The noise and fumes encourage those who can afford to do so to live somewhere else.

In the UK, the Chief Medical Officer's 2022 report estimated that air pollution in England leads to the premature deaths of between 26,000 and 38,000 people a year, with, in addition, many more suffering from avoidable chronic ill health because of it.[12] Analysis of official UK government data reveals that air pollution disproportionately impacts lower-income and more deprived areas, and particularly affects neighbourhoods with higher ethnic-minority populations. In a study using 2020 data, Friends of the Earth found that half of the neighbourhoods that had very high air pollution were in the bottom 30 per cent of the most deprived neighbourhoods.[13] On top of this, they found that nearly half (44 to 47 per cent) of the population in neighbourhoods with very high air pollution were people of colour, meaning that members of this group were over three times more likely to live in a highly polluted area than white people.

There was (and is) also a disparity between those producing pollution and those most impacted by it. For example, households in neighbourhoods with very high air pollution were up to three times less likely to own a car than those in the least polluted areas. Friends of the Earth also observed how dirty air creates a massive burden on the health service and businesses, with illness and lost work due to outdoor air pollution costing about £20 billion a year in England.

When it comes to mortality and ill health, pinning down the contribution of pollution on a case-by-case basis is often tricky when other factors are also involved, perhaps explaining why there hasn't been a greater popular demand for action to fix it, but every now and again there is a case that proves the general point. In 2020, for the first time in the UK, and possibly the entire world, a death certificate recorded air pollution as the cause of death. On 16 December that year, Southwark Coroner's Court in London found that air pollution 'made a material contribution' to the death of nine-year-old Ella Adoo-Kissi-Debrah.[14] She lived close to a major road in South London and died following an asthma attack. The ruling that determined air pollution to be a contributory factor cited excessive exposure to nitrogen dioxide from vehicle engines as the reason.

Ella's mother Rosamund subsequently campaigned for a new clean air and human rights law, which, in November 2022, was introduced to Parliament for debate, bringing pollution and social inequalities closer together on the political agenda.[15] London Assembly member Leonie Cooper, who supported Rosamund's campaign, made the connection clear between the inequalities inherent in how pollution affected different groups. 'Disproportionately, those dying are from Outer London, and are from Black, Asian and Minority Ethnic backgrounds,' she observed.[16]

Official data back this up, and the specific tragic case of Ella gave it a human face. As to why there isn't still more of an outcry on this subject, even when that tricky connection between cause and effect is increasingly made manifest, is in part down to the difference between who is being polluted and who is doing the polluting, and to the relative power and influence they each have. The fact that air pollution has been much higher on the agenda in the British capital during recent years is helped by having Sadiq Khan as London mayor. As a man of British Asian origin, unlike many of his predecessors, he has a different take on the priority that needs to be attached to taking action to cut air pollution.

RESOURCE EXTRACTION

Pollution from industries which process natural resources and the effects arising from the consumption of the products they make, including the fuels that power vehicles, often impact disproportionately on disadvantaged groups. The same can often be said about the extraction of these resources. Demand for minerals such as copper and iron, timber from natural forests, wild fish, fossil fuels and land to grow crops, such as the vast plantations of oil palms that increasingly replace rainforests across tropical countries, causes not only environmental damage but also a range of social impacts.[17,18,19] These resources are often found in remote areas, including places inhabited by Indigenous societies. It is these people who are at the front line of the environmentally destructive extractive processes that sustain our economy. The process began centuries ago and continues today.

The loss of tropical rainforests across South Asia, from India to New Guinea, has meant that the once continuous canopy of dense green is now fragmented and, in most formerly forested regions, largely gone. As the forest has been cleared, social conflicts have arisen alongside serious environmental challenges. One struggle that I became aware of during the late 1980s was that of the Penan people. This tribe lived (and some still do) in the rainforests of northern Borneo, in the Malaysian state of Sarawak, where for decades timber companies have been going ever deeper into their forest territories to extract trees for wood. The palm oil companies have followed them, replacing the depleted forests with the vast monoculture plantations that feed the lucrative market for this valuable vegetable oil, which ends up not only in food, but also in various consumer goods, including toothpaste and cosmetics.

Some 35 years after my first forays into the campaigns mounted by the Indigenous people of north Borneo, I met with their representatives in London to discuss their continuing struggle. Komeok Joe leads the Penan organization Keruan and helps organize the resistance against the destruction of their rainforests. With a slight wiry frame, piercing eyes and thick dark hair, he was probably about my age, although he looked younger and was much fitter than me. He knew the history of the campaign very well and told me about blockades that he and other Penan had been placing in logging roads to prevent the companies getting access to the forests, going back to the 1970s.

The tribe's decades-long fight slowed forest loss and drew attention to their struggle. 'This is the only way we can survive,' he told me. But even after years fighting back, much of the forest is now gone. The companies take a first cut of trees and come back for a second later on, and then entirely clear the natural forest. 'After they finish the timber they start palm oil plantations,' Komeok Joe explained. He had come to the UK to seek support for their movement to protect the forests, along with his colleague Celine Lim, a Kayan who leads the resistance for other Indigenous groups that live in Borneo's forests – namely, the Kenyah and Saban – via their grassroots organization SAVE Rivers.

Celine told me about the ways in which the fightback from the Indigenous people had led the Malaysian Timber Council to launch a labelling scheme for sustainable wood. 'The market now demands sustainable timber. Because of that the Sarawak government went towards that direction. But the question is whether there has been a change since the sustainable scheme has taken root.' The answer, she told me, was that it hadn't. She explained that there was even one area where the new scheme for sustainable timber was approved during the pandemic and the communities didn't know anything about it. 'We did a lot more digging in understanding how the sustainable scheme would work, but at the end of the day there is no difference between before this sustainable scheme and after the sustainable scheme.'

Labelling wood as 'sustainable' of course helps forestry companies, wood suppliers and retailers answer awkward questions from customers seeking to implement environmental policies, including those running shops in the UK, but if they are not based on reality, and do not include the views of affected communities, such schemes are at risk of being dismissed as greenwash. In this case, a large and powerful timber-exporting company has clashed with Indigenous groups seeking to protect their forests, with the labelling scheme driven more by sustaining business as usual than by preserving local culture, biodiversity and the forests' carbon stocks.[20]

Considering the limitations she saw in this consumer labelling system, I asked Celine if the real question was more about who had the power to decide what happened in the forest. She said that native customary rights are recognized in Malaysia, which should give some level of control over the land to the people who live there. She added, though, that 'when it comes to the work on the ground there are again a lot of loopholes'. This includes the quality of the maps that set out who has what rights where. 'Even though there was a map, it was a 1958 map,' she said, pointing out that the nomadic Penans have very different concepts of land and who uses it. They were confronted with a *fait accompli* by people with far more power and influence than them. She reasoned that if Indigenous people had clear legal rights over their ancestral territories, there would be no need for a sustainable wood label. 'We have kept our forests intact for generations and that says a lot, right.'

The question of who has control of the forest is key, not whether a labelling scheme should be introduced or not. The Indigenous groups have fought on via multiple strategies: they have used the law (deficient though it often is), sought international support (hence their visit to London), protested with physical resistance (including the blockades on logging roads), and organized information campaigns, including about the sustainable timber scheme.[21] One response to the latter approach from Samling Plywood, a major player in the extraction of timber from Indigenous lands, was to silence dissenting voices via legal action which sought to prevent SAVE Rivers and others from making what the company regarded as defamatory statements about the timber-labelling scheme and its deficiencies. On top of this censorial injunction, the company demanded an apology, and made a claim for damages equivalent to 45 times SAVE Rivers' annual income.[22]

During the early 1990s, I became involved with another Indigenous struggle, on the part of tribes living in the Amazon rainforests of Ecuador. The oil industry is among the economic interests invading their lands. One leader who has been involved with that struggle for more than 30 years is Domingo Peas, an Achuar leader from the region where northern Ecuador borders Peru, and president of the Amazon Sacred Headwaters Alliance.[23] This region is known among ecologists as the Tumbesian area of endemism and it is one of the biologically richest and most unique areas on Earth. I became aware of it while working at BirdLife International. We were mapping such areas to enable us to target conservation resources where they would make most difference – for example, in plans for new national parks. In the case of the Tumbesian region, 58 endemic (restricted-range) species of birds are found there, making it the third richest 'biological hotspot' on Earth. In the wake of deforestation, however, some 23 of those unique birds are threatened with extinction. Other species are in similar peril.

I spoke with Domingo about his work and the pressures he is resisting. He told me how his community's focus on the conservation of their rainforest is a solution to the crisis we all face. His people know very well about global heating, loss of species and deforestation, and why it is critical to halt and reverse these trends. They take a broad view about what is needed to halt the destruction, and know the changes must go far beyond what they must do on the ground to protect the forest where they live.

He highlighted the need for more awareness among the people who are using the resources extracted from the forests. 'The people who are consuming oil or other commodities need to know that these things are coming from rainforests, and they have to know that these things are coming at a cost of destruction of those ecosystems.' He urged people to pay more attention. 'Consumers should be aware, they should have consciousness, not just about the Amazon, but from wherever the products that they're consuming come from, what ecosystems are paying the price, and to make sure that they're eliminating the supply side of where these commodities are coming from, and if they're coming from this destruction we have to stop that.' He also cautioned against being fooled by false claims. 'Cacao is supposed to be an alternative economy,' he said 'but in many cases it is grown as a monoculture that's replacing the rainforests. So even the product that's supposed to be a beneficial product is a harmful product.'

As with the Indigenous campaigners in Borneo, it comes down to an asymmetry of power. Those making the most consequential decisions include governments in distant capital cities, executives in the boardrooms of major companies and those of us who are the end consumers of the resources extracted at great ecological cost. From Borneo to Brazil, from India to Indonesia, and from the Congo Basin to Canada, Indigenous societies who've lived close to nature on their ancestral territories for millennia have been subject to external pressures that have caused them in many instances to lose their customary and ancestral lands, and their culture and identity. Some have been enslaved and others completely wiped out by external interests seeking to exploit them or their lands, and often both. The inequality of power that permitted this has not only been a social and cultural tragedy, and in some cases genocidal, it is also a major contributor to the ecological catastrophe that is rapidly unfolding on our planet.

The replacement of the nature-focused worldview of Indigenous societies with the resource-focused worldview of profit-driven companies, and the economic growth sought by the nation-states that facilitate this, explain in many instances the difference between the areas of intact nature that remain compared with those places where it has been destroyed. And in the process of extracting the resources used to make lives more convenient in distant cities, another form of environmental injustice is perpetrated. For when the products made from the resources extracted from mines, forests and soils are disposed of and turned into waste, a further set of unevenly distributed consequences follows on.

BURNING PLATFORM

Having run out of space to create new waste landfills across many densely populated areas, and having encountered limitations as to what can be recycled (because of technical challenges, industry resistance, lack of facilities and consumer rejection), there has during recent decades been an increased focus on waste incineration, including technologies that convert waste into electricity, and sometimes heat too. I've come across multiple local campaigns opposing the siting of such plants next to where people live, not least because of fears over the health risks posed by the pollution plume from their furnaces.[24]

Cancer-causing dioxins and various toxic heavy metals were among the compounds that concerned campaigners. Over time improvements have been made and toxic releases have been reduced with more advanced scrubbers and filters, but even with cleaned-up flue gases there are good reasons why many still object to having an incinerator located close by. These include concerns about litter, fears that local trees and hedges will be festooned with plastic bags and that there will be increased vehicle traffic with frequent movements of refuse trucks, and, of course, the pervasive sweet and sickly smell of our consumer detritus that wafts across nearby neighbourhoods as it awaits combustion.

Considering what we have seen so far, you might not be surprised to learn that such plants tend not to be sited in better-off areas. Like other polluting infrastructure, they are more likely to be built in poorer areas. Take the one located just outside the Cornish village of St Dennis, near the town of St Austell. St Dennis has, in common with many other English rural and coastal towns, high levels of social deprivation; it was also the place selected to host Cornwall's new energy from waste plant.[25,26] It is striking that out of the whole county, the new facility was built in a place ranking among the 20 per cent most deprived areas.[27]

As potential places to locate this new piece of infrastructure were being considered, vocal opposition drove it away from more affluent communities, contributing to a pattern repeated across the UK. According to analysis published in 2020 by Unearthed, a research group in Greenpeace, waste incinerators are three times more likely to be built in the most deprived neighbourhoods than in the most affluent ones.[28] The research also found that people of colour were over-represented in the areas where incinerators were sited.

The quantity and types of waste generated through our high levels of material consumption are visible manifestations of the unsustainable trajectory upon which we are embarked. As might be expected, that level of consumption and waste is disproportionate and reveals another layer of inequality. Of all the waste arising in the world each year, about 34 per cent of it is generated by high-income countries, which, between them, have about 16 per cent of the world's population.[29] When it comes to the waste that flows through consumer societies, one of the responses that most of us will be aware of, and support, is recycling. It is generally a more sustainable option than incineration, a way of preserving and reusing resources, rather than destroying them. Unfortunately, the consequences of even this apparently benign and positive approach can have negative impacts, which fall harder on the poorer sections of societies. Take plastic recycling, which is increasingly driven by rising consumer concerns about the damaging effects of plastic waste, especially on marine life.

Many people have sought to reduce their consumption of single-use plastics, governments have taken steps to discourage it, and companies have cut their reliance on such materials.

At the same time, there have been attempts to recycle more plastic, by shifting to recyclable packaging and raising consumer awareness about the need to put waste plastics in the correct bin. Assuming used plastic does indeed go in the right bin, its journey afterwards can be less positive than those of us rinsing our yogurt pots would like. The waste is often shipped from developed countries to poorer ones, where there are insufficient facilities to process it.[30]

Of the 2.5 million tonnes of plastic packaging waste collected for recycling in the UK in 2021, over 60 per cent of it was shipped abroad.[31,32,33] When it arrives in the countries where it is meant to be processed, sometimes the facilities to properly deal with it are lacking, leading to a range of damaging consequences, including health-threatening pollution. Turkey has been the main destination for British plastic, causing what one parliamentary committee described as 'irreversible and shocking' environmental and human health impacts.

Clothes are also shipped abroad, including to low-income countries, for recycling and reuse. Two investigations (conducted by ABC and Greenpeace Germany) looking into used-clothes supply chains found that some 40 per cent of what is imported into these nations from the Global North is deemed unusable at the other end of its journey and so is landfilled, dumped by the side of roads or burned.[34,35] Some 70 per cent of such garments are made from synthetic fibres and thus are also plastic waste (although not categorized as such).[36] When these materials are burned, toxic pollution is released, including cancer-causing dioxins.

In Indonesia, where the UK sends some of its waste for 'recycling', material is mainly burned or openly dumped. Even that destined for actual recycling can be problematic. This includes bales of used paper from the UK, Europe, Australia and New Zealand that are typically contaminated with approximately two to ten per cent of plastic, but during recent years, that proportion has increased significantly.

The paper companies have no facilities to process the plastic, so it is sold for fuel. Some of the toxic pollution from burning plastics finishes up in the human food supply. One investigation looking into the fate of these hidden plastics found high levels of dioxins in free-range chicken eggs collected near a tofu factory in East Java that burns plastics for fuel.[37] Dioxin was found in the eggs at concentrations close to those detected near the hotspot in Bien Hoa, Vietnam that is considered one of the most dioxin-contaminated locations on Earth following the use of the notorious defoliant Agent Orange by the US military during the Vietnam War in the 1960s and 1970s.

Where it works, recycling can indeed be part of the solution, but in some circumstances it can act as a cloak for one more dimension of environmental injustice. Imports of plastic waste pollute the local environments of poor communities, on top of the waste that those communities already generate themselves, which they are also poorly equipped to deal with. In the case of Turkey, the capacity exists to collect and process only about 20 per of the country's own plastic waste, so imported material diminishes that proportion still further. In Indonesia the picture is worse still, with only about five per cent of its own waste being recycled. At the global level, around nine per cent of plastic waste is recycled, with nearly a fifth incinerated and about half landfilled.[38] Of the remainder (22 per cent) that ends up floating in the ocean, caught in tree branches, or converted into pollution by being burned in the open air, some will have been sent to developing countries from the richer ones, as part of their 'recycling' effort.

After decades of target setting and at times genuine progress, it is still the case that resource extraction, processing and waste disposal hit the disadvantaged hardest. They tend to live downwind, downhill and downstream of pollution and on the front line of where environmental blight meets people. That this remains a modern problem is underlined by the US Environmental Protection Agency's 2020 report, which pointed out how even in the world's largest economy, '[l]ow-income, minority, tribal, and indigenous communities are more likely to be impacted by environmental hazards and more likely to live near contaminated lands'.[39]

While questions of environmental injustice that emerged as a political idea as long ago as the 1970s have proved resistant to solutions, a whole new dimension has now come to the fore. More than four decades since *Blade Runner* depicted a world blighted by environmental disaster, aspects of the dystopian future it set out seem more plausible with each passing year, manifest not least in the rapidly rising trend of global heating. This huge environmental challenge touches all aspects of society and is accompanied by injustices on an epic scale.

5

Hot House

The linear economy that extracts resources, emits pollution and wrecks ecosystems (with consequences that fall disproportionately on disadvantaged people) is also changing the composition of the Earth's atmosphere. The global heating that comes in its wake hits vulnerable people first and hardest. One country on the front line is the Maldives, an archipelago of small coral islands scattered across a vast area of the central Indian Ocean. It is widely referred to as a small island state, but the people who live there sometimes prefer to call it a large ocean state.

Like most people who go to the Maldives, I arrived there via Malé airport. Approaching the Maldives' capital in a Boeing 777 over the vastness of the Indian Ocean, I spotted the runway from the plane window, looking as though it nearly filled the tiny island it served. It seemed a precariously small target for such a big plane, set as it was just above sea level and fringed by boulders to protect it from ocean waves. The airport is the city's most important physical connection with the outside world, and as we got closer I could see that Malé's urban area was spread across two further islands, one just over a mile across and another, thinner than the other, about three miles long. Home to about 140,000 people, they were crammed with colourful apartment blocks, mosques, shops, port facilities and streets.

My first impressions of this unique island country included a sense of isolation, as well as vulnerability, which, considering that it is the lowest-lying country on Earth, with its highest point only 1.5 metres above sea level, I discovered was not unfounded. From the international airport I transferred to the nearby Noovilu Seaplane Terminal, where I boarded a flight to one of the even more remote islands, Kunfunadhoo Island, located about 100 miles to the north. It was late 2014 and I was en route to a meeting of environmentalists.

As with many other environmental gatherings, the challenging irony that the participants arrived by plane was not lost on those who'd chosen to come. The calculation for me was down to whether I thought our discussions might move things on in a way that would make the carbon emissions a price worth paying for a bigger goal. High on the agenda were preparations for COP21 in Paris, scheduled to take place a year later. Everyone knew that would be a make-or-break meeting, and those gathered in the Maldives had it high on the agenda, including colleagues from China who were working hard to improve the chances of success from that side of the political jigsaw. Our meeting also raised questions about the fact that an island nation dependent for nearly a third of its GDP on luxury tourism needed to maintain large-scale air links to keep that going.[1]

Flying over deep blue sea to an island less than a mile long and a few hundred yards across, I could see the patterns of coral reefs below the water revealing rich life living in the sparkling seas, out of which small points of land here and there peeked out, the tips of the huge submarine mountain range that this tiny country is created from.

That journey underlined for me why the Maldives is considered one of the most dispersed countries on Earth, comprised of 1,192 coral islands grouped in a double chain of 26 atolls stretching some 541 miles north to south and 81 miles east to west. Putting the idea of a large ocean country into perspective, some 35,000 square miles of the Maldives are ocean and just 115 are land. This far-flung country is home to a total of about 557,000 people, spread across

200 or so inhabited islands. Very low lying, with an economy dependent on its ecological assets, the Maldives is highly exposed to the effects of global heating.

The impacts of that are already being felt and threaten to deliver blows that will strike at the very heart of Maldivian society, and indeed threaten the country's existence. The predicament faced by this vulnerable nation is one reason why the goal to limit global warming to below two degrees centigrade, and as close as possible to 1.5 degrees, compared with the pre-industrial temperature, was in the end adopted at COP21. At that meeting, at the insistence of the small island countries, that target was taken on as the global goal because if heating wasn't limited to about 1.5 degrees, they'd disappear.

Sea level is already going up. It has been for decades, presently by about three and a half millimetres per year, due to ice melt on land and thermal expansion of the ocean. That rate might rapidly increase if major bodies of ice become unstable, as some in Antarctica are showing signs of becoming. At the same time, the warming of the sea impacts critical ecosystems, such as coral reefs, which are additionally damaged by the acidification of the ocean caused by the absorption of carbon dioxide from the air.[2] As a result of these stresses, the colourful reefs become bleached, taking on a ghostly pallor as the corals expel their symbiotic algae and die.

As the delicate balance of the reef is tipped over, so effects on a wide range of species follow, including those that are of great importance to people for food. And as the corals die back, so do the tourism-based businesses that rely on them. On top of these climate-related consequences, there are others, such as the damage caused by more frequent and extreme storms. One person who thinks about all this a great deal is Aminath Shauna, who, from 2021 to 2023, was the Maldives' Minister of Environment, Climate Change and Technology.

I met Shauna at a meeting in London hosted by King Charles to follow up from the successful Montreal talks on biodiversity that took place at the end of 2022, when a new Global Biodiversity Framework had been adopted. I spoke to her afterwards about the situation faced by her country. She told me, 'The Maldives is one of the lowest-lying nations in the world, which makes us extremely vulnerable to any amount of sea-level rise, any fraction of a degree of warming as well. We are already seeing the impacts of climate change here.' She went on to describe the wide range of impacts now in play. 'The rain pattern has changed. Our coral reefs are bleaching at an alarming rate. We are already seeing that the frequency of coastal flooding has increased, and also, when it does rain, the rain pattern has changed so much the number of islands that get flooded has increased as well.' The main reason why the Maldives is so important for biodiversity is because of those coral reefs, which is why she was there at that gathering with the King.

'We have about three per cent of the world's coral reefs in the Maldives,' the minister told me, 'and we are the seventh largest coral reef system in the world.' She explained how the 1998 El Niño led to the loss of nearly all the Maldives' coral reefs, and how, after they had largely bounced back, they were hit again in 2016, at which point about three quarters of these hugely biodiverse systems were destroyed once more, turning the vibrant wonders of the sea into white deserts, leaving only the skeletons of the dead corals, eerily monochrome and still. I saw the aftermath of a bleaching event on the reefs at Kunfunadhoo Island, which, as part of the Baa Atoll Biosphere Reserve, are recognized as globally important.

Located in the central western part of the Maldives, the atoll supports one of the largest groups of coral reefs in the Indian Ocean, and acts as a stepping stone for the transport of the planktonic larvae of reef organisms from the western and eastern Indian Ocean. When I snorkelled there, it was just recovering from a bleaching event, with fish returning to the few areas where the coral was once more growing – but it was soon to be hit by yet another heat event. 'Today, what is happening is every year is a warm year for us, which is making it impossible for the reefs to adjust and to adapt to these changes,' Shauna explained. 'So it's a very scary and a very unpredictable situation for us here because the coral reefs is what we depend on for everything, for protection, for food and for income as well.' But the damage being caused to the biodiverse reefs is unfortunately not the only thing that is causing climate change to impact so seriously on the economy of the Maldives.

For while the country's ability to earn income from tourists visiting to dive on its incredible reefs is being compromised by climate change, the impacts of climate change are also increasing costs. Shauna described how more and more of the Maldives' limited government budget is consumed on dealing with progressive climatic change, including the more frequent disasters that are already being endured, and on preparing for future ones that will certainly follow. The government is bracing itself to deal with continuing cost escalation as the country's hundreds of islands need more coastal protection.

One key area causing particular concern relates to water security. 'Every island in the Maldives has also run out of fresh drinking water because it has been contaminated due to saltwater intrusion,' the minister told me. As a result, the government is building desalination plants on each of the inhabited islands, and this, alongside flood and coastal protection, is now demanding more and more of the nation's financial resources. 'It comes to about 30 per cent of our domestic budget, which means we have to divert critical funds from education, from the health sector and from other developmental needs.'

Climate change is having a political impact too. 'The Maldives is an emerging democracy, so we want to continue to consolidate democracy, which means we need to work on spending more on institution building,' explained Shauna. 'If we are going to have to continue to spend more on disaster risk, resilience and building more shore protection and flood prevention and water resources, that's really taken critical funds which we would otherwise be using for these developmental needs.'

How all this will pan out in our complex, fast-changing world is not clear, but Shauna is worried. 'It's going to be extremely challenging. I cannot imagine a world that is more than 1.5 degrees because these are the changes that we are already seeing at 1.1 degrees of warming.' Her words present a vivid case in point as to how the impact of climate change feeds through to poorer members of society who depend more directly on government expenditure to educate their children and look after their healthcare needs. The money can only be spent once, and if it's going on sea defences and desalination plants, it's not going on schools or medical scanners.

In late 2024 it was confirmed that global average temperature had entered what is widely regarded as the danger zone of over 1.5 degrees of global heating, beyond which it is expected a series of serious impacts will likely occur. While for many politicians in some countries this seems to be merely a matter of passing concern, in the Maldives those in elected office know it is a life-or-death threshold. 'We saw in the recent IPCC report that global temperatures could go beyond 1.5 even before 2030,' said Shauna. 'I just fail to understand why the world cannot come together on addressing the climate emergency that we face today. It's not the lack of knowledge. It's not the lack of technology. The real issue is this is all about politics. If there's political will, you can really make bold decisions to keep it under 1.5.'

We will return to this point about politics later, and about how countries like the Maldives which feel the worst impacts of climate change are among those that have contributed least to causing it.

The small island states are, of course, not alone in being under such threat, for while it is the case that people living on low-lying islands and around vulnerable coasts are often highly exposed to the impacts of climate change, this is far from the only context in which socially disadvantaged people find themselves particularly vulnerable to the effects of global heating.

EXPOSED

Most people in the Maldives are not wealthy, but neither are they among the poorest of the poor. I met some of the people who carry that dubious accolade a few years ago while researching the social and economic dimensions of deforestation in the forest frontier landscapes of the West African country of Côte d'Ivoire

(Ivory Coast). I went there in 2016 to conduct research for my book *Rainforest*, and met people living on the fringes of the only natural forests that remained in what had become a largely deforested country.

As it has done for hundreds of millions of others living close to the edge in their day-to-day existence, poverty has rendered rural communities there highly exposed to climatic shifts, including severe drought. One of the main sources of livelihood for rural people across West Africa is cocoa. Many of the farmers who produce this key ingredient for the global chocolate brands work small plots of land, some barely producing enough to make a sufficient living to feed themselves and their families. When it doesn't rain, the fall in crop yield can mean the difference between having enough to eat and going hungry.

Farmers working there told me about the recent intense heat and dry spells the likes of which they'd never seen before, and how this had impacted their livelihoods. Global heating was one thing, and another was the deforestation that made the effects of climatic changes worse, as the towering rain clouds that once bubbled up over the rainforests each day dwindled as tree cover declined, becoming less and less reliable.

Their predicament is shared by hundreds of millions of Africans producing food on small rain-fed plots, who find themselves at the front line of climate-change impacts. In common with the cocoa farmers of Côte d'Ivoire, most of them produce not only to sell, but also to sustain themselves and their families. Being hit by drought, cyclones or a pest or disease attack thus has a double impact, on income and on direct food supply, which in turn is made worse by lack of money, especially when scarcity causes prices to rise.

For communities where there is little by way of a social safety net, these consequences that become more frequent and extreme with climate change can be catastrophic. And the impacts are chronic as well as acute, bringing long-term effects of malnutrition, such as stunted growth and vitamin deficiencies, as well as starvation in the short term. These longer-term effects on food production can affect educational attainment, which in turn leads to intergenerational disadvantage, with implications for development more broadly.[3,4,5] There are, unfortunately, more and more examples of how these connections work in practice.

Take the pastoralists living on the far side of Africa from those cocoa growers, in parts of Ethiopia, Kenya and Somalia, who, during the early 2020s, suffered through a sustained drought following the failure of rains across four consecutive wet seasons. In 2023, large areas of Somalia faced the prospect of famine.[6] For people raising camels, goats and sheep, rain is vital to sustain pasture, and if that fails, animals die and livelihoods collapse, thereby removing the main means of securing income and thus money to buy food and other essentials.[7] Elderly people with the longest memories of living in that part of Africa have recently described these droughts as of unprecedented severity, including one in 2011 that killed a quarter of a million people. A further drought in 2016 and 2017 led to the mass death of livestock, leaving the pastoral economy weakened and vulnerable to the severe dry period that then followed during the early 2020s.

In April 2023, United Nations Secretary-General António Guterres visited Mogadishu to see for himself the crisis playing out on that climate-change front line.[6] Accompanied by the UN Humanitarian Coordinator for Somalia, the world's most senior diplomat met some of the Somalis affected by the crisis. One family had travelled 105 kilometres on foot and by donkey cart to seek refuge in the Baidoa camp in 2022 after all their livestock perished from drought. A second family had done the same after all their animals died, travelling 70 kilometres to seek aid.

Their stories are but two among those of 8.25 million people then needing lifesaving assistance from the international community because of the most recent climate shock, involving several consecutive years of poor rain. Of those 8.25 million, 3.8 million were internally displaced, with the other 4.5 million or so suffering from the effects of acute food insecurity. Almost two million children were severely malnourished, with eight million lacking access to adequate water and sanitation. Two thirds of those in the drought-affected areas had no access to essential healthcare.

A humanitarian response plan set out in 2023 to meet the crisis estimated that Somalia needed about 2.6 billion dollars, but had secured only about 15 per cent of that sum.

The climate-impacted smallholders like the pastoralists in Somalia and the cocoa growers of Côte d'Ivoire are among a vast global army of food producers who run about 84 per cent of all the world's 608 million farms.[8] Between them they cultivate and graze about 12 per cent of the Earth's agricultural land. The International Fund for Agricultural Development estimates that smallholder farms in the developing world support the livelihoods of almost two billion people.[9] These small-scale food producers are mostly in tropical developing countries, and they make up not only most of those growing food, but also about half of those who do not have enough of it. That proportion rises to more like three quarters in Africa.[10,11] Given the evidently huge overlaps between poverty, food security and vulnerability, it is clear that if there is to be a realistic prospect for reaching global targets on these key themes of the SDGs, then how the world deals with climate-change impacts on this highly exposed group will be a big determinant of success, or lack thereof.

Climate-change models suggest that the kind of real-world events observed during recent decades are set to become more serious, with the expectation that even moderate temperature increase will have negative consequences for rice, wheat and maize production. These crops are the mainstay cereals for smallholders, and it is expected that in a warming world the effect of drought and flood and the impact of pest and disease outbreaks will lead to the increased frequency of poor yields and livestock mortality.[12,13]

Many of the regions where these impacts are expected to be most severe are tropical countries with large rural populations of poor smallholder farmers. Most of these farms lack irrigation and, at the end of dusty tracks, are often far from sources of disaster assistance and even if the farmers had money, they often have little access to shops and are frequently pushed to the edge of food security, even in the good times. On top of that, much of their land is degraded, and, with depleted organic matter, their soils are vulnerable to drought and erosion. A shock can tip them over the edge and into destitution. Those of us living in western countries see these people every now and again, when they become the subjects of a harrowing news story or the reason for a new fundraising appeal.

In California about 80 per cent of arable land is irrigated, whereas in Niger, Burkina Faso and Chad, this figure is less than one per cent.[14] In the USA more than 90 per cent of farmers have crop insurance to cover losses in the event of extreme weather, whereas only 15 per cent of farmers in India benefit from such cover, ten per cent in China, and under one per cent in Malawi and most other low-income countries. Compared with their counterparts in rich countries, farmers from the poorest half of the planet's population live overwhelmingly in countries that are considered not only the most exposed to climate change, but also much more vulnerable.

Voices from those most vulnerable nations find it hard to break past the status quo. At the COP28 climate negotiations held in Dubai in 2023, where there was an intense discussion about the merits of adopting a target to phase out fossil fuels, the status quo was defended with vigour. Considering that the use of such energy sources is the principal reason for the climate changes now causing such damage to regions like East Africa, there was a serious scientific rationale for such an undertaking. With strong representation of fossil fuel interests there, however, seen in the thousands of attendees from the oil, gas and coal industries, as well as vocal petro-states, including Saudi Arabia and Russia, there was intense pressure to weaken this ambition. So numerous were these representatives that they exceeded the number of people there to represent the countries regarded as the ten most vulnerable, including Somalia, Chad, Tonga and Sudan.[15]

THE GENDER DIMENSION

In those countries most exposed to climate risks, not everyone is equally vulnerable. Women living in the rural parts of developing countries are especially hard hit. They are often most directly dependent on natural resources for their livelihoods, do most of the agricultural work and are responsible for collecting water and fuel. Climate change can thus affect their lives profoundly. For instance, increased climate variability is making agriculture more unpredictable, while continuing land degradation and desertification exacerbates the domestic fuel crisis. In addition to these factors in many societies, including rural agricultural ones, women tend to have less money than men, magnifying the ways in which poverty increases their vulnerability to climatic shocks.

The scale of the disparity is reflected in the estimate that women make up 80 per cent of climate refugees globally, which in turn is one factor leading to the UN Special Rapporteur on violence against women and girls concluding that 'climate change is the most consequential threat multiplier for women and girls'.[16,17] It is, of course, not only the physical impacts of global heating, including displacement, falling disproportionately hard on women that shape such a situation – their social status is significant in this regard too.

This point is highlighted in a World Bank study which found that where women have a lower socioeconomic status than men they tend to die in greater numbers during and immediately after disasters.[18] With such findings in mind, it is evidently the case that it is not only drought, fire, flood and pests that are issues, but also the social and economic contexts of the people affected, and, once again, a determining factor is the extent to which vulnerable people can influence decisions.

One study into the impact that extreme weather linked to El Niño had on poor Peruvian women provides an example of how social marginalization and inequality in decision-making can lead to worse social outcomes.[19] El Niño is a natural climatic phenomenon that occurs on a more-or-less five-yearly cycle, when the cold Humboldt Current that flows north from Antarctica along the coast of Chile and Peru is replaced with a warmer southern-flowing current coming from the opposite direction. This change from a polar to a tropical current raises sea temperature and causes heavy rainfall, floods and landslides in some areas, and drought in others. The severity of El Niño's impact varies from year to year and from place to place, but in 1997–1998 it was particularly severe in parts of Peru.

Over 100,000 homes were either damaged or destroyed by floods and landslides, affecting around half a million people, with three quarters of those affected from rural areas. Although El Niño is a natural phenomenon rather than a result of human-induced climate change, the frequency of extreme weather linked with it is increasing because of climate change.

Researchers looked into the impact on people living in the uplands of the northern coastal Peruvian province of Piura. This mainly rural part of the country is dominated by small-scale farming which is, for the people who live there, the principal source of food and income. Despite this, it suffers from food insecurity, exacerbated by the impact of repeated El Niño episodes. This is in part because there's no official programme to support small farmers, and in the absence of that, there has been a tendency for farmers to move to urban areas in search of work. Because they've struggled to make a living from farming, locals have been forced to seek additional income from other activities, including plundering woodlands, which in turn makes the risk of landslides worse. The damage to crops caused by El Niño led to household incomes falling, and when the weather was most severe and some communities were cut off, not only did they have little food, they also had less money to buy the available food, which, because of the bad weather, was more expensive.

Women were found to be especially vulnerable in the wake of these events, due to lower access to education, specialist technical assistance and healthcare and lack of control over decisions concerning land and farming, which was their ultimate source of income. And when there was less food (and even when there was enough), women were found to have less of it, with women and children disproportionately exposed to the risk of malnutrition and disease epidemics, including acute respiratory and diarrhoeal infections, malaria, dengue and cholera, all of which increased significantly during El Niño.

Although women were disadvantaged, they were nonetheless left in charge of the households, as men moved out of farming areas into the coastal valleys and cities in search of work. Women were, however, not recognized by the main community organizations, which were led largely by men, and their increased household burden meant they couldn't seek paid work. Against this backdrop of social disadvantage, national policy nonetheless favoured the development of large-scale agricultural export industries, including cotton and rice (rather than small-scale enterprises to supply local markets), with rural upland communities largely excluded from development priorities. This meant the women living there had to resort to various self-generated survival strategies to cope with the problems linked with the extreme weather.

Whereas it was thought by the Peruvian government and international agencies that economic growth would be generated by exports in turn funded by large quantities of financial capital ploughed into the agri-business and other sectors, researchers investigating the plight of the farming communities concluded that it was social capital that was lacking. Social capital is the web of networks and connections, relationships, local institutions and knowledge of the land and what it can do that, if effectively harnessed, helps communities cope with shocks and setbacks. Since the 1997–98 El Niño-related weather event, and as a result of women's groups effectively seeking change, there's been an increase in women's participation in planning and decision-making, and as a result, disaster-prevention strategies have been strengthened.

This is important to know in a world that tends not to focus on fostering growth in social capital in order to create resilience to climate shocks, preferring instead to spend on concrete and drains in often vain attempts to defend communities from extreme rainfall. The tendency is also to invest in large-scale projects that most visibly boost economic growth, rather than building the resilience of the communities at the sharp end of change.

CITY CLIMATE

It's not only poor rural communities living close to the land that are increasingly hard hit by global heating – urban ones are too, including those affected by extreme hot weather. Now and again the threat posed by heat hits the headlines, as it did at the end of 2022 with revelations about the deaths of thousands of workers employed to prepare for the Qatar World Cup.[20] Due to incomplete record keeping and reporting, it is not possible to say exactly why so many Nepalese, Indian, Bangladeshi and other migrant workers lost their lives (an estimated 6,700 of them in all) during the ten years of preparations for the tournament, but the extreme heat in Qatar during the summer months was a major factor.

Qatar is a very hot and dry corner of the world and is indeed among the countries with the least rainfall and highest average temperatures. As the world warms, it's getting more extreme still, as the ocean surrounding the peninsula upon which it sits progressively overheats. The Gulf region in which Qatar is located is warming more quickly than many other regions, and it is expected that by 2070 much of it will have conditions beyond the tolerance of humans. That part of the world is, of course, not alone in seeing increasingly serious impacts from hotter conditions. Indeed, about 40 per cent of the world's population lives in climates where the average daytime maximum temperature for most of the year exceeds 30 degrees centigrade.[21] Most living in these countries are poorer members of the global community and many of them are engaged in physical work, especially outdoor labour, and are thus particularly susceptible to the effects of high temperatures.

Poorer workers in developing countries are among the most vulnerable to this aspect of climate change, and due to their limited choices, they are least able to avoid its consequences. These people also live in substandard housing and their physically demanding jobs are rewarded by output, with no or very little insurance or employment alternatives.[22]

As a result, they are more vulnerable than better-off workers (including those in air-conditioned offices) to the effects of heatstroke, cardiovascular diseases, exhaustion, fainting, digestive problems, kidney diseases and, in extreme cases, death. The situation is worsened by limited knowledge about occupational health and safety issues, with that lack of awareness sometimes exacerbated by a lack of interest. Another dimension to the impacts of extreme heat is the extent to which poor women are likely to bear the brunt of health problems caused by it.

Globally, the cities which experience the most extreme heat are all in developing nations. One analysis revealed how the number of days in which individual city dwellers were exposed to extreme heat went from 40 billion in 1983 to nearly three times that by 2016 (119 billion).[23] This increase is partly down to the rising urban population globally, but a significant part of it is also due to climatic heating. The worst-hit city according to this analysis was Dhaka, the fast-growing capital of Bangladesh, with others across Asia and the Arabian Peninsula showing similar trends.

A heatwave of record-breaking intensity that hit Asia in 2024 saw thermometers from India to the Philippines reporting temperatures in excess of 40 degrees centigrade, with 50 degrees recorded in several countries, including India, now the most populous country on Earth. This led to deadly conditions for those undertaking manual work outside. Birds and mammals died, overwhelmed by the heat. In Bangladesh all schools were closed. It was a further glimpse of the ever more extreme world we are creating through continued greenhouse gas emissions, and of the consequences that come with that. Baked into the impacts are several social inequalities.

One study from Brazil (pertaining to the city of Belo Horizonte) found that the ethnic-minority population, low-income residents and the elderly were statistically significantly more associated with heat risk.[24] This study also observed how these groups lived in the city centre, unlike comparable urban communities in developed countries, where the poor tend to live in the periphery.

With many developing countries in sub-Saharan Africa, the Middle East and South Asia in a phase of rapid ongoing urbanization, the challenge is set to become worse as the climate warms further. With temperature and humidity extremes in many locations already regularly exceeding what the human body can safely withstand, this juxtaposition of trends might ultimately challenge the economic viability of many major cities. Also playing into this is the fact that poverty reduction in many urban areas ultimately rests on increased labour productivity, with the poor moving to urban areas in the first place to secure employment and, they hope, better lives. As elevated temperatures become increasingly linked with decreased economic output, it may be that some profound consequences emerge in the wake of urban heating in developing countries, as hundreds of millions of people are unable to realize the economic gains that should have accompanied their labour.

Severe city heat also hits some developed countries, and there too consequences fall unequally across society. For example, analysis of UK National Health Service data on emergency hospital admissions between 2001 and 2012 revealed that in periods of significant heat (and, indeed, cold as well) it was the old, the young and those from low-income backgrounds who were more likely to be admitted to hospital.[25] Investigations in the USA to compare heat-island effects with social groupings found that people of colour were much more likely to live in areas most affected by extreme heat than non-Hispanic whites in all but six out of the 175 largest urbanized areas in the continental US.[26] A comparable pattern emerged in relation to those people living below the poverty line. Research in the UK conducted by the University of Manchester for Friends of the Earth reached similar conclusions, revealing how people of colour are, compared with white people, four times more likely to live in areas at risk of dangerous levels of heat.[27]

In the USA on average the poorest ten per cent of urban neighbourhoods were 2.2 degrees centigrade hotter than the wealthiest ten per cent on days with extreme heat and during average summer weather.[28] The difference was as high as 3.3 degrees centigrade in some urban areas in California, such as Palm Springs. Similar disparities have been detected in the UK, where the Index of Multiple Deprivation has been found to correlate with higher temperatures.[29] The reasons for these kinds of disparities are varied and, as we shall see in the next chapter, linked in part with the proportion of trees and green spaces in different neighbourhoods. In some places, they are also linked to deliberate spatial planning policies. This is one reason why in the USA more deaths are caused by excessive heat than by other kinds of extreme weather, and extreme heat is also associated with non-fatal illnesses, including heatstroke and dehydration. These health impacts fall most heavily on poorer and marginalized members of society, which is also the case when it comes to other kinds of extreme conditions.

KATRINA STORMS ASHORE

In late August 2005, a violent Category 5 hurricane formed over the Atlantic Ocean off the east coast of the USA. When it smashed into the coast of Louisiana at the mouth of the mighty Mississippi River it inundated parts of New Orleans. The devastation left in the wake of this epic weather system revealed how natural disasters are not great levellers that impact people equally, but very often disproportionately hit those least able to cope. This is not to say that storms and other extreme weather events seek out weaker victims, but neither do they arrive in social or economic vacuums. And when storms like Katrina crash through defences that prove flimsy when set against the power of nature, they can exacerbate inequalities, bringing into sharp focus social vulnerabilities and inadequacies in the ability of even the most advanced societies to respond to climate change.

Climate science is getting better at determining whether extreme events are linked to human-induced global heating, but, as with many other major weather events, it is not possible to say for certain that this storm was caused by human-derived changes to climate. It is, however, correct to say that more frequent extreme weather is part of an emerging and documented pattern connected to the rise in global average temperature, in turn caused by the progressive build-up of greenhouse gases in the atmosphere. The additional energy trapped in the system leads to warmer air, carrying more moisture, propelled along on winds that can be stronger than would otherwise be the case. One study that looked at the impact of Katrina estimated that flood elevations would have been 15 to 60 per cent lower at the start of the twentieth century than those observed when the storm hit in 2005.

The level of flood elevation is linked with the extreme low pressure that accompanies storms like Katrina. This sucks up the ocean surface to create a 'tidal surge', lifting the water beneath the weather system so that when it hits land, inundation by the sea causes more devastation on top of that caused by the rain and wind. In the case of Katrina, the ten or so inches of rain that fell in parts of Louisiana were accompanied by a maximum wind speed of 175 miles per hour, with the storm surge lifting the sea to about 15 feet (4.5 metres) higher than normal, causing extensive flooding, property damage and loss of life.

Most of New Orleans is very close to sea level and some of it is below, with the city rendered more vulnerable to severe weather because of the progressive degradation of the freshwater marshes that once spanned the delta of the Mississippi. This ecological degradation was due to seawater penetration arising from the construction of canals, thus removing what was once a layer of natural protection between sea and city. Manmade barriers in the form of a series of levees proved inadequate to hold back the rising sea and the hurricane-force winds, which swept aside not only that defensive infrastructure, but also the veneer of US society, revealing beneath a highly stratified social order in which the poverty that so often accompanies Black American communities also made them more vulnerable.

New Orleans used to be a prosperous city, and in the 1840s it was one of America's richest. Its economy was based on natural resources and fuelled by slavery, and it was for a time the leading marketplace in the country for buying and selling people. Over time it declined and became one of the nation's poorest cities, although the wealth gap between Black and white remained deep. When Katrina hit, the rate of poverty among Black people was triple that of the white population. An especially disadvantaged area that took the full force of Katrina was New Orleans' Lower Ninth Ward. It's called the Lower Ninth because it is physically low lying compared with the rest of the city, and far down the river by the sea. So when, at 10 a.m. on 29 August 2005, the levee wall protecting this part of the city was overwhelmed, it was one of the first parts of New Orleans to be flooded.

A wall of water rushed through neighbourhoods, smashing homes to pieces. In the aftermath of the storm, mud-smothered cars, refrigerators and household goods lay jumbled in the streets. As the world's media began to cover the tragedy, it was images of the city's residents stranded on rooftops and crowded in appalling conditions in the then Louisiana Superdome that summed up Katrina's human face. People found themselves marooned without assistance, food or water for days, and most of them were not only poor, they were also Black.

While many of the worst excesses of overt racism in America have diminished over decades of painful change, it is evidently the case that lingering injustices are exposed by environmental threats, including toxic pollution and the effects of climate change. To many white people, these continuing differences in people's experiences are not obvious, and thus not a priority to fix. Barack Obama, who was an Illinois senator at the time of the hurricane, said on the floor of the Senate that the response to the disaster was not 'evidence of active malice' but more the result of 'a continuation of passive indifference'. His points were subsequently borne out by surveys conducted by the Pew Research Center in 2015, ten years after the storm hit.

New Orleans residents were asked about the federal government's response to the storm, and only 19 per cent of Black people said it was excellent or good, compared with 41 per cent of white people.[30] Three times as many white people (31 per cent) as Black people (11 per cent) said President George W. Bush did all he could to get relief going as quickly as possible. Some 71 per cent of Black people said that the response to the hurricane showed that racial inequality remained a major problem in the USA, while most white people (56 per cent) said this was not a particularly important lesson from Katrina.

History will continue to remember Katrina as the costliest natural calamity in US history (until a bigger disaster strikes) and to record the tragic loss of more than 1,000 (mostly Black) lives and recall the misery that came in its wake. What the world might also care to remember is how it was a powerful example of the ways in which social inequalities render some more vulnerable to environmental change than they might otherwise be, with the greatest costs of all hitting the poorest, and in this case also predominantly Black, communities the hardest. It is very important to note, however, that as the effects of global heating hit the most vulnerable, there will also be knock-on consequences that affect the better-off.

CLIMATE, CONFLICT, TERRORISM, REFUGEES AND SECURITY

As conditions become more difficult for those living on the front lines of climate-change impacts, they have been inclined to leave their homes to seek a new life in more benign conditions. During my travels in Côte d'Ivoire, I spoke to people in remote villages who, following the decline in incomes resulting from drought, told me of people they knew who'd left to travel across the Sahara Desert to Libya in the hope of finding a way to join a boat crossing to Europe. The consequences of that perilous journey are increasingly in the news, as boats sink, people drown and controversies break out in European countries over what to do about migrants. Major political consequences can follow.

One of the most potent images deployed to encourage voters to support Brexit was a poster depicting queues of refugees beneath the slogan 'Breaking Point', accompanied by words which proclaimed 'we must break free of the EU and take back control of our borders'. During the summer of 2016, the lines of refugees building up on European borders were mostly Syrians, fleeing a civil war. What was not publicized too much at the time was the extent to which the internal Syrian conflict was a further social manifestation of extreme weather, and likely also of human-induced global heating.

In a paper published in 2015, climate researcher Colin Kelley and his colleagues set out a series of connections that related extreme weather caused by climate change to social unrest and civil war in Syria.[31] They presented data revealing how the 2007–10 drought that preceded the country's decline into civil unrest and conflict was the most severe ever recorded, causing crop failure and the mass migration of farming families from the countryside to urban areas.

They analysed weather records going back a century and results from climate modelling to conclude that the severe three-year drought was made two to three times more likely by human induced climate change than would be otherwise expected through natural variability alone. The authors said it was notable that three of the four most severe multiyear droughts ever recorded had occurred during the 25 years prior to when they conducted their analysis and concluded that 'human influences on the climate system are implicated in the current Syrian conflict'. The connections these researchers explored paint a chilling picture of the risks that are ramping up with more extreme conditions.

The story began with the impact of drought on water security, and the subsequent effect on farming as soil moisture dropped, crops withered, pastures shrivelled and animals died. This in turn caused as many as 1.5 million people to move from the farmed countryside to the towns and cities. Critical in causing this movement of people was the effect of ill-conceived policies that for years had damaged farmed landscapes. These included policies from the government of President Hafez al-Assad (the father of President Bashar al-Assad, who was in power when the civil war started) that aimed to increase agricultural output through more irrigation projects and subsidies for diesel.

These policies put pressure on already stressed water resources at a time when drought was becoming more frequent, causing the groundwater which supported much of the country's farming to become depleted. Over-abstraction of water caused rivers and irrigation canals to dry up. Across wide areas, farming could no longer continue. When drought hit, this depletion of the groundwater meant that the farming in the country's north-eastern breadbasket region, where some two thirds of Syria's crop yields were produced, collapsed. When the water evaporated, so did farm incomes, and when wheat had to be imported, there was in 2007 and 2008 an unprecedented rise in Syrian food prices. In a single year, the price of wheat, rice and feed more than doubled.

Rural poverty and hunger sparked mass migration to the peripheries of Syria's cities, which were already pressured following up to 1.5 million Iraqi refugees arriving in the wake of the Second Gulf War. This meant that the urban population of Syria increased dramatically, and fast, growing by more than 50 per cent in under a decade, placing housing, services and infrastructure under massive strain. This then exacerbated factors often cited as contributing to the unrest that ultimately exploded into civil war, including, the authors said, 'unemployment, corruption, and rampant inequality'.

The civil war, of course, had ramifications that ran far beyond the borders of Syria, creating a refugee crisis that not only added to political tensions in Europe, but also prepared the ground for the spread of the Islamic State terrorist movement (ISIS). Such is the web of interconnections that shape events in the modern world, with, in this case, social inequality combining with climate change to create an explosive mix.

When the brutal regime of Bashar al-Assad finally came to its abrupt end in December 2024, there was little reflection or recollection in the coverage that followed as to the underlying causes of the unrest that led to 13 years of misery. The depredations visited on the Syrian people created a powder keg of resentment, and the spark that finally ignited the explosion was in part linked with wider pressures and trends, and which in our ever more stressed and volatile world it is important not to lose sight of. And what happened in Syria is not the only conflict that during recent times has been linked with the consequences of global heating.

General Richard Nugee is a British Army officer with internationally recognized expertise in climate change and security. He is chief adviser to the all-party parliamentary group on climate and security and the University of Oxford's Climate Change and (In)Security Project. His insights are derived from a variety of perspectives: he has led a review on how the British military can deliver net-zero emissions, and he experienced the very sharp end of where environmental change meets strategic risk with two tours of duty in Afghanistan, in 2006–07 and 2013–14. He told me about the linkages between climate change and violence in that country and how, despite most Afghans opposing rule by the

Taliban, it nonetheless managed to recruit many new members. Nugee discovered that one reason for this was failing agriculture.

Farmers there grew crops for local markets and opium poppies for heroin production. In both cases, they depended on the gentle annual melt of snow and ice in the Hindu Kush mountains that dominate much of the country, releasing water that would irrigate the land and sustain at least one and sometimes two harvests per year. Nugee told me how climatic heating interrupted this once reliable process: 'What we found in the Hindu Kush, and particularly in the northeast of Afghanistan, is as you got much higher temperatures, snow would melt really quickly and then there'd be nothing else to melt in the summer. So instead of irrigation you get a flood and then you don't get irrigation you get a drought and then it snows in the winter, and then you get the cycle repeated, which is flood followed by drought.'

Not only were changing patterns of ice melt linked with impacts on crop production, they were also reflected in rising levels of violence. This was in turn down to destitute farmers having no option but to join the Taliban. 'The Taliban understood what was going on. They're not stupid,' Nugee told me. 'They paid five dollars to a farmer a day to join the Taliban. They would join as foot soldiers specifically to gain money so that they could feed their families, and they got a bonus if they killed one of us. Fifty dollars for every western soldier killed was what we understood.'

Syria and Afghanistan are unfortunately not the only places where climatic shifts have been linked with subsequent violence. General Nugee told me about how the 2023 coup in Niger has been called a 'climate coup' by African analysts. They traced links between the nine major droughts and five major floods in the last 20 years and water shortages and food crises, and the subsequent rural poverty and dissatisfaction, which created opportunities for the non-democratic seizure of power.[32] 'You get what we're seeing in lots of other places and desperation to move to an alternative,' said Nugee. 'That desperation is manna from heaven for non-state actors wishing to overthrow the government.'

He noted that this same story of political instability caused by rural economies failing because of climate impacts is repeating across other areas of Africa's ecologically fragile Sahel region. 'You're seeing the same in Mali as well. Firepower rules, the person with the biggest gun rules. You get a feedback loop of violence, corruption, displacement, all exacerbated and created by climate change, which eventually leads to a coup.' He described the political dynamic in play. 'In Mali, the government forces were seen as the enemy. The non-state actors have nothing to lose because they don't have any responsibility. They can promise the world, that they will sort everything out.' He told me how farmers were also actively targeted for recruitment by ISIS in Iraq and how Boko Haram had done the same in Somalia, where those suffering from frequent flood–drought cycles have ceased to have a livelihood. 'What you end up with is people desperate for money to be able to feed their families,' he said.

Nugee points out, as other researchers do too, that climate change is not the only factor in play. It comes against a backdrop of rural poverty, rapid population growth, economic dissatisfaction, weak governance and few services. Reducing the risk of violence therefore requires action to alleviate poverty, tackle entrenched inequality and help people reduce their vulnerability via sustainable development. That in turn invites questions for the international community as well as for the countries on the front line. From what he has seen, Nugee concludes that sustainable agriculture must be the priority, helping fragile rural communities cope with what are now inevitable changes. 'If you provide grain, that's no good, right? If you provide money, that's even less good. What you've got to do is provide the expertise to be able to improve the agriculture of the local environment.'

Many people living in places where there is a failing rural economy, and, worse still, where conflict has resulted, will wish to move. Many are already doing so, with most of those displaced by climate-related changes moving within the borders of their own country. Some, including those with more money, leave their home nation to travel internationally, and this too can contribute to political tensions. We already see some of that in the UK, and considering the scale and pace of climate change, we could face circumstances in the coming years that make the recent political controversy of small boats appear minor by comparison.

From the rural pastoralists whose animals have died to the farmers who can no longer grow enough to make a living and the small islands where the freshwater supply is swamped by the sea, millions of people are expected to be on the move as climate impacts reap their toll. According to the Internal Displacement Monitoring Centre (IDMC), between 2008 and 2015 an annual average of 21.5 million people were forcibly displaced by what they call 'weather-related sudden onset hazards', including floods, storms, wildfires and extreme temperatures.[33] Many more are fleeing their homes in the wake of slower-moving crises, such as coastal erosion linked with sea-level rise. The IDMC estimates that '71.1 million people were living in internal displacement worldwide at the end of 2022, a 20 per cent increase in a year and the highest number ever recorded. The Democratic Republic of Congo, Nigeria, Afghanistan, Ethiopia and Yemen had the highest numbers of acutely food-insecure people in 2022. They were also home to more than 26 million IDPs [internally displaced persons], over a third of the global total.'[34]

What is already a very serious challenge is set to get a great deal worse. It will be those with the least ability to adapt and the least means to ride out the storm who will be hit hardest. Many of them, especially the young men, will also move first, and the disadvantage that caused them to be vulnerable in the first place will become a problem for everyone as climate refugees clog the immigration systems of the countries they head to. In some cases, the armed forces of western countries will be drawn into conflict, leading to loss of life and huge expense, as was the experience in Afghanistan.

Profound environmental injustices are not restricted to how the most vulnerable are hit by the consequences of global heating, toxic pollution and resource extraction – they also include unequal access to healthy green environments. That is the theme of the next chapter.

6

Human Nature

We humans have an innate affinity with the natural world. This is perhaps not surprising, considering how we are as much a part of the natural order as trees and birds, with every second of our existence spent relying on nature, from the oxygenation of our blood through to the provision of our food and the replenishment of our fresh water. Even within our bodies there are multitudes of other species: the more than 100 trillion individual organisms living in our guts include between 300 and 500 different kinds of bacteria carrying some two million genes.[1,2]

On top of this intimate personal dependence on what nature does for us, we have for hundreds of millennia been close to nature's diversity, cycles and seasons, with our brains and physiology tuned to work in settings where connectedness with how our world works was part of our day-to-day experience. In our modern context, however, it is fair to say that those of us who have more contact with nature tend to appreciate what it does more than those who have become detached from it.

In 1965, when I was four and a half years old, our family moved from a rented flat in an old Victorian house to a 1920s council house with a big, long garden. It was in the Oxford suburb of Cowley, and the garden had wild corners with clumps of bramble. My mother hated its unruly, wild demeanour, but I loved it. There were nesting blackbirds in it, and a wet marshy area at the very bottom next to a small stream where frogs could be found. There were old willow trees by that stream that at night attracted tawny owls. Swifts nested under the eaves in the roof. I became obsessed with these mysterious long-distance travellers, sleeping close to them in the little back bedroom where at night I could hear them shuffling about in their nests.

Next to the stream between the willows was a row of old hawthorns where bullfinches were among the birds that nested. Butterflies flew across the garden, and some stopped to lay eggs, producing caterpillars. Big bumblebees lumbered between the springtime blossoms on old apple and plum trees. For me as a child with a nascent interest in nature, it was an outdoor classroom that sparked intense fascination, compelling me to travel ever further afield to look for more wildlife.

Abandoned industrial works lay beyond the stream at the bottom of the garden, and, with boys from along the road, I'd crawl through holes in an old fence to explore the untamed area beyond, looking for voles, moths and birds' nests. At about seven years old, I was given a bicycle. With horizons widened by pedal power, I embarked on nature forays to other abandoned 'wastelands'. One was an old farm swallowed by the suburbs decades before, but which remained undeveloped. It was known as 'The Tres' because a sign that once declared 'Trespassers Will Be Prosecuted' had been hit by rocks and general decay to the point where only the first four letters remained.

Among the thickets of brambles and hawthorns that smothered the crumbling walls of The Tres were thrushes, finches and warblers. Summer visiting birds included turtle doves, with their flimsy twig nests to be found in the hawthorn thickets, from where they also delivered their dreamy purring song. I spent many hours there, interacting with the wildness of the place, and taking home mental images to compare with those printed in the Ladybird and Observer's books about wildlife. As time went on, I became interested in trees, butterflies, fish, reptiles and fossils, travelling further to more abandoned places, including old railway sidings, quarries and gravel pits, where there were not only rocky remains containing traces of creatures that had lived in the distant past, but also grass snakes, slow worms, toads, newts and more. The more I looked, the more I found, and the more I found, the more I looked. Very importantly, I discovered that human knowledge of the natural world was not complete. One day I found a male blackbird incubating eggs, yet my bird book said that only females did that. It was an exception that fuelled a sense of discovery that motivated me to engage more with the nature around me.

These kinds of experiences deepened my curiosity and laid the foundations for what was to become a lifetime passion. It was a youthful journey that, as it did for many other wildlife enthusiasts of my baby-boomer generation, led to an interest in conservation. This evolution from merely looking at nature to also being concerned about its disappearance was stoked by my day-to-day experiences as those wild patches were progressively filled in with new developments of houses, offices and roads.

Today we call many of these remaining wild urban areas 'brownfield sites', and such places remain the preferred target for built development, in turn progressively driving down the remaining wildlife populations that have clung on in and around the towns and cities where most of us live. There is green space remaining, or course, in many urban areas, although unlike those informal wild oases that I explored as a child, the public parks that most people use for outdoor enjoyment are rather sterilized of wildlife, with mown grass, planted trees and little by way of self-willed nature.

As a child, I felt disquiet at the loss of wild spaces, but I can see today that I felt it more than most because I had become so immersed in the natural world. Recent research has revealed, however, why the progressive disconnection with nature that comes with such decline is a major issue for all of society, rather than just being something of concern to those of us lucky enough to have had opportunities to become so engaged with it. It affects us all because our connection with nature runs so deep that our need and desire to connect with it is increasingly seen as a hard-wired instinct, known as biophilia.³ Some believe that an affinity with greenness has been baked into us humans through our evolutionary exposure to drought, leading us to welcome it as a sign of better times. But while the threads that bind people to the biosphere are fundamental, over time they've become progressively frayed, especially as more and more people have come to live in towns and cities, where in many cases less and less wildness remains.

As the world population exploded from one billion in the early nineteenth century to more than eight billion today, the proportion living in urban areas has also grown inexorably. This progressive shift to living in built environments has marked a period of massive change for humankind. Throughout our entire history, until about 200 years ago, the vast majority of people lived close to nature. We have been in daily connection with the natural world for about 99 per cent of the period since modern humans first evolved on the savannahs of East Africa.

People's livelihoods, and, indeed, their survival, depended on having detailed knowledge of nature to enable them to engage in hunting and gathering. Later, other nature-based economies, including farming, forest industries and fishing, required similarly detailed knowledge of the natural world to succeed. Administration, manufacturing and other non-nature-based roles were the exception, not the norm, as they are now. The shift from rural to urban living went hand in hand with industrialization. When that process got going, about 30 per cent of the population in England lived in urban areas, rising today to well over 80 per cent. It was not only work that drew more and more people to the towns and cities, it was also the inns, the fashion, the theatres, the social networks and, for some, the desire to escape from the relative conservatism of rural areas.

The shift from livelihoods derived from the land and sea to those based in cities and their factories and offices has been a steady global trend. In 2007, for the first time in human history, more people lived in urban settings than in rural ones, and by then the population was on the way to seven billion.⁴ It is expected that by 2050, when the United Nations estimates the total number of people will have grown to about 9.8 billion (about 1.8 billion more than in 2023), more than two thirds of us (an estimated 68 per cent) will be urban dwellers.

GREEN AND HEALTHY LANDS

Accommodating the massive urban growth anticipated over the coming decades will require vast quantities of cement, steel, glass and wood, and, to service the increased population, huge infrastructure expansion, including water supply and sanitation, power, heat, cooling and transport, and, of course, enormous supplies of food to sustain the non-farming city dwellers.

While the environmental impact of urbanization has tended to focus on the loss of undeveloped areas, other factors of concern include the vast resources needed to create more urban areas, the pollution caused by vehicles and the energy consumption of households and workplaces. Increasingly, more subtle and hitherto underappreciated consequences of urban environments are garnering interest. And while it might be self-evident that access to green space is good for people, sometimes it is necessary for even obvious conclusions like that to have data and science behind them before they are taken seriously.

Following years of evidence gathering it is now possible to say with scientific certainty that access to green and blue (that is rivers, lakes and oceans) spaces is good for human well-being in relation to both physical and psychological health. Even short-term exposure to forests, urban parks, gardens and other semi-natural environments reduces stress and symptoms of depression, combats attention fatigue, improves self-esteem, lifts the mood and creates an overall sense of improved well-being.[5,6] Part of the benefit comes from how natural environments trigger fascination, helping to foster mental restoration through switching the brain off from the stresses of everyday pressures and preoccupations.[7]

Access to outdoor spaces tends to increase physical activity as well, thereby improving overall public health through having a positive impact on obesity and related conditions, such as type 2 diabetes.[8,9,10,11] Living in areas with a high proportion of green space and landscape diversity has been associated with reduced mortality from respiratory illness, cardiovascular conditions and cancer, and with improved respiratory and mental health; the beneficial effects are apparent even among people who live up to five kilometres from green areas.

The positive effects of natural environments on people's well-being have been found in every stage of life, including in pre-natal development, with the greenness of mothers' neighbourhoods correlated with a positive effect on the birth weight of infants.[12] Childhood exposure to green space has been associated with lower risk of schizophrenia, while the greenness of residential areas where young people grow up has been linked with reduced prevalence of allergies in children, and has a positive impact on the blood pressure of adolescents.[13,14,15,16] Some of these effects are not only linked with the greenness of what we can see and experience through sound, smell and touch, they are also operating at a microbiological scale, with exposure to microbes in early life creating positive effects on the immune system and the prevalence of chronic inflammatory diseases.[17]

Access to nature-based experiences brings benefits for people held in prison, and it helps children concentrate during school lessons.[18,19] One famous early study from the USA found that patients recovering from surgery did so faster and with the need for fewer painkillers when they could see green areas from their hospital beds, compared with patients who could not.[20]

On the flip side for young people are consequences including emotional, cognitive and physical difficulties that arise from a lack of interaction with nature during early life, especially due to limited access in urban environments. A set of mental disorders linked with a lack of connection with the natural world has been called nature-deficit disorder by writer Richard Louv in his 2005 book *Last Child in the Woods*. The term was coined to describe the human costs that come with alienation from nature, and although it is not a medical diagnosis, it underscores a growing problem that previously didn't have a name. Since it was first put forward it has become seen as a phenomenon that applies not only to children, but more widely across societies.

In addition to these broad associations between access to green spaces and human well-being, there is a developing body of research that points to biological diversity having a bearing on the value of our green spaces too. While this remains an underexplored area, some linkages have already been identified. One relates to plant variety and the extent to which the structural diversity that comes with a habitat comprising ground flora, shrubs and trees can be more effective than a green environment comprised of a mowed lawn at combatting air pollution, and thus more beneficial in shielding people from its effects.[21]

More natural and diverse environments will also generally be expected to have a greater diversity of micro-organisms, and thus a more beneficial effect on people's immune systems.[22] Studies looking into microbiological environments and the health of six-to-thirteen-year-old children living in Bavaria in southern Germany found that the prevalence of asthma and allergies was significantly lower among those growing up on family farms compared with the wider population of young people.[23]

There is evidence that the diversity of larger organisms in natural environments is also linked with the beneficial effects of being in those environments, and that the pleasure people derive from being in ecologically diverse places increases with their knowledge of the names of the species that they see there. Studies have found how songbirds are especially appreciated, with evidence that exposure to a variety of birds reduces depression, anxiety and stress.[24,25] The positive effect is linked not only with what people can see, but also with the singing and other sounds made by such creatures.

The evidence of positive associations comes from studies conducted across the world, from the USA to Australia and from the UK to Japan. Woodlands, rivers, coasts and hilly landscapes with expansive views have all been found to touch people in positive ways. Urban trees lift property values and in one US study have even been found to be correlated with lower levels of domestic violence.[26,27] Another study found that increased urban green space was associated with lower property crime in all but one of the 301 US cities the researchers looked at, and with a reduction in violent crime in 289 of them.[28]

In addition to the positive well-being impacts that come from time spent in more natural areas and the ways in which vegetation can cut down air pollution, there are studies suggesting that wildlife populations might help to protect people from infectious diseases. For instance, one study from the US state of Louisiana found that transmission of the deadly West Nile virus from mosquitoes to people went down when there was greater bird diversity. This seems to be because the mosquitoes, which are the vectors of that disease, prefer to feed on birds, so when there are more birds they have less need to suck blood from humans, thereby reducing the risk of the virus being transmitted. Evidence for a comparable dilution effect has been found among other pathogens and parasites too.

A MANDATE FOR NATURE

On top of these health-enhancing and health-protecting contributions of nature, access to natural areas reinforces connections that are vital for creating and sustaining the public mandate that is needed to enable politicians and company executives to come forward with effective environmental programmes. The more experience and knowledge people have of the natural world, the more likely it is there will be clear demand for such programmes from a willing and informed public motivated by their personal concern for nature and wild places.

If most people don't know much about nature, they are less likely to care about it, and then those running businesses and governments will find it all the harder to do what is needed, even if they themselves care (and, considering how a lot of them evidently don't, this point is even more critical). Sir David Attenborough is one among a number of well-known environmental figures who has pointed to the importance of there being a broad popular mandate for action to protect and restore nature, which in turn will rely on people having access to natural areas where they can experience nature's wonders for themselves.

High-quality TV programmes such as *Wild Isles*, *Planet Earth*, *Blue Planet* and others have had a major impact on public awareness, but they often portray the rare and special, the remote and unusual – aspects of nature that many of us will never know or see from direct experience. There is research to confirm that such programmes can help foster pro-environmental action by the public, yet knowing nature from personal experience is qualitatively different to what we can absorb from TV, not least because of all those health and well-being benefits already mentioned.[29]

Market democracies are to a very large extent driven by popular opinion, majority votes and public demand for change. With that in mind, it is clear that nature must be available for everyone to enjoy directly on a day-to-day basis in and around the neighbourhoods where they live. This in turn requires focusing on nature recovery in towns and cities, not only in the countryside, and especially in those areas that are presently most deprived of green and blue spaces.

A strong and growing body of evidence reveals how feeling connected to nature can stimulate feelings of personal responsibility and lead to pro-environmental behaviours, such as recycling, cycling rather than driving, avoiding unnecessary waste and not dropping litter.[30,31] This positive cycle, in which spending time in nature leads to a greater propensity to look after it, often starts in early years, with overwhelming evidence showing how children who spend regular time in the natural world are more likely to exhibit pro-environmental behaviour as adults.[32,33,34,35] It is a self-reinforcing synergy too, with positive experiences of nature in childhood leading to a greater likelihood of spending time in nature as an adult, which in turn reinforces pro-environmental behaviour. There is also evidence to strongly suggest that it is not simply a matter of being exposed to green spaces, but that our emotional connection with these spaces is important. The evidence for this positive relationship comes from around the world.

One study from Italy found that adults were more likely to make sustainable food choices if they'd had childhood nature experiences, such as gardening or observing wildlife: the greater their connectedness to the natural world, the higher their frequency of making sustainable food purchases.[36] Connectedness to nature was found to be an important driver of ecologically friendly behaviour among American children aged nine to twelve, with connectedness explaining 69 per cent of ecological behaviour and just two per cent of such behaviour resulting from environmental knowledge.[37] Work from Spain supports such an emphasis, indicating that long-term exposure to nature through attending summer camps is an effective way of promoting children's emotional affinity with the natural world, helping to foster ecological beliefs and support their intentions to adopt environmentally friendly behaviours.[38] Further studies from Germany and Taiwan, among other countries, confirm the importance of childhood recreation in nature, often with family members, as a predictor of future behaviours that protect the environment.[39] Work from developing countries reveals similar connections. A study from Brazil with a sample of 224 young adults found that greater contact with nature during childhood was associated with greater contact as an adult, which in turn was positively associated with connectedness to nature and pro-environmental behaviour.[40]

A study in England found positive relationships between nature appreciation and pro-environmental behaviour across socio-demographic groups, whereby the more individuals visited nature for recreation the more they appreciated it and the more pro-environmental behaviour they reported.[41] Positive associations between such behaviours and close access to neighbourhood green spaces and coastal proximity were found across the board, in both high- and low-income households, thereby revealing – contrary to some popular mythology – that pro-environmental behaviour is not automatically a middle-class tendency.

Apart from anything else, these kinds of studies might lead us to question the level of effort expended on seeking to influence people through the communication of facts compared with that put into making real connections with nature. Knowledge is very important, but research reveals it is not enough on its own to cause people to take action. Meaningful, positive experiences of nature that lead to emotional affinity with healthy natural surroundings are more important. In other words, sporadic and occasional contact with the natural world is insufficient to instil in children and adults the curiosity, wonder and connection required for nature to become a meaningful part of their lives.

These findings are backed by intuitive feelings that are in turn reflected in art and literature. William Wordsworth is but one of many writers inspired by nature whose appreciation of the power of the relationship between nature and humanity was a fundamental inspiration. In his poem *The Prelude*, finally published in 1850 following decades of reflection and refinement, he wrote beautifully about nature as a 'disciplinary force', proclaiming its importance to a person's intellectual and spiritual development, showing how a love of nature can lead to a love of humankind, and detailing how a poet's mind must grow through mutual consciousness of and communion with nature. What he might write today about the relationships between people and nature among those living in areas bereft of natural areas is an interesting thought experiment.

I wonder if such a poet might reach conclusions similar to those of the social scientists – namely, that habitual experiences in nature as part of daily life are what is generally needed to shape harmonious outcomes between humankind and the rest of creation that sustains us. Whether they are reached through art or science, however, the same conclusions can be drawn – that in order to harness popular backing for the protection of nature and environmental recovery, access to nature cannot only be available to a minority of people. If we are to shift outcomes in democracies, support will need to come from right across society, including from among groups that are presently excluded from regular access to attractive natural areas.

The fact that there is a need for an inclusive, society-wide approach logically flows from the research, and so does the importance of early-years engagement between people and nature. For while connectedness to nature can be a driver of pro-environmental behaviour throughout life, it is positive direct experiences in nature during childhood and role models close to the child who care for nature that are the two most important factors driving adults to choose to take action to benefit the environment.

It is perhaps unsurprising, therefore, that most people who choose a life in which they demonstrate commitment to the environment report significant experiences with nature in their childhood. For most of these individuals, the natural habitats were accessible for unstructured exploration and discovery nearly every day when they were children. I can say for sure that it was my experiences of nature at a young age, whether on family holidays, in our garden or in the various wild corners that I explored, that were the main reason I devoted my life to ecological advocacy. And I continue to spend a great deal of time outside in natural areas, which undoubtedly contributes to my personal well-being.

As wilder areas have diminished in number and extent, so too have the opportunities for people to experience wildlife themselves. Looking around the places where I grew up today, six decades later, I am sad to see that much of what first inspired me and my friends has gone. It is no wonder that wildlife species that were once common, such as thrushes, hedgehogs and various butterflies, are now much scarcer, including in urban environments.

This is lamentable for many reasons, including the very practical one of people's health and well-being, and the pressing urgency of creating the popular demand needed to shift societies into sustainable trajectories before it is too late. As is the case with other environmental changes I have described in relation to pollution, resource extraction and climate change, there are major inequalities in how the issues are manifested among different groups of people.

UNEQUAL ACCESS

Given the benefits to well-being that come with ready access to nature, it is perhaps not surprising that homes with bigger gardens, that are closer to natural areas, that are adjacent to riverside walks, or that have pleasant views of green and blue areas are more desirable. They are therefore more expensive than those that have little or none of the above.[42,43,44] To this extent, one basic link between equality and access is down to where people can afford to live, with less wealthy people more likely to be excluded from greener environments. It is also noticeable how it is the better-off people living in the most expensive neighbourhoods who are most successful in fighting off developments that will lead to the loss of the green spaces near them. Incinerators, roads and other developments are more likely to be sited in poorer areas, where green and wilder spaces are also more likely to be lost and, indeed, blighted by different kinds of pollution.

In England, as wild corners have been progressively eaten away by continuing urbanization, there have been various attempts to encourage more people to use National Parks, National Landscapes, National Trails and other areas officially designated for public enjoyment in beautiful environments. This is welcome, important and necessary, but of course visiting these places requires transport, knowledge of where to go and the confidence to explore unfamiliar, faraway places. People who are less well-off and therefore less likely to be able to afford to live in green areas are also less likely to have a car, or to be able to afford long train journeys to reach accessible areas of countryside.[45] Many of those living in low-income households also often have less leisure time as they hold down multiple jobs and work shifts on weekends, when their children are not at school.

On top of these barriers, there are also sometimes fearful perceptions of the outdoors to contend with, some of which are linked to cultural differences, a subject I will return to shortly in relation to race and access to the outdoors. These are among the factors that help to explain the very considerable inequalities in access to green spaces, and thus the unequal distribution of the benefits that can come with spending time outdoors.

Evidence from across Europe reveals that less green space is available in less well-off urban neighbourhoods compared with higher-income ones.[46] Communities with a low average income, low levels of educational attainment and high unemployment rates tend to have access to smaller areas of green space than those with high incomes, high levels of educational attainment and low unemployment rates. Children from lower socioeconomic backgrounds in Germany were found to be disadvantaged in terms of access to urban green spaces compared with children from wealthier families.[47] In cities across Central and Eastern Europe, properties in areas with more green space tended to be more expensive, a tendency seen in Poland and in Debrecen, Hungary, where new upmarket neighbourhoods have more green space than older housing estates inhabited by lower-income residents.[48,49]

Not only do people on lower incomes tend to have limited access to green space, such space is also often of lower quality than that in wealthier neighbourhoods, in turn reducing people's motivation to use it. One study found that in socioeconomically disadvantaged neighbourhoods in Helsinki, Berlin, Bucharest and Lisbon, urban parks had less diverse facilities and vegetation than those in wealthier city areas.[50] In the Netherlands, green areas in poorer neighbourhoods were found by researchers to be less aesthetically pleasing than those in wealthier neighbourhoods.[51] In Porto, Portugal, in addition to offering fewer amenities, green spaces accessible to populations of lower socioeconomic status have more signs of damage that in turn give rise to more safety concerns than are felt in wealthier neighbourhoods.[52] One UK study found that although people who live in deprived urban areas both recognized and appreciated the value of local green spaces, they tended not to use the closest and most convenient ones because they were often of poor quality and felt unsafe. That study revealed that less than one per cent of people living in social housing reported using the green spaces on their estates.[53]

The now rich research literature on unequal access to green areas includes a study from Sweden which found that people with higher incomes have more access to nearby green and blue areas within walking distance of their homes.[54] In the city of Łódź, Poland, a study found that two thirds (67 per cent) of children who belonged to the poor-income-related status group had very low visible greenery along their walks from home to school.[55] In Porto, Portugal, the average distance it takes households to reach green spaces increases with neighbourhood deprivation. In the Netherlands, the availability of green space within 250 metres of people's homes was less in neighbourhoods of lower socioeconomic status.

Outside Europe, one study looking at five US cities found a striking lack of access to safe parks for those on low incomes, while a review of five Australian cities also found highly unequal access to green spaces: in four of those cities, lower-income neighbourhoods had less green space than better-off ones.[56,57] In the developing countries of the Global South a similar pattern of low access to green space broadly prevails.[58] And around the edge of those cities, in large informal settlements of shanties and favelas, there are even bigger overall wealth inequalities than are seen in Europe and North America. Cities of the Global South are also often more polluted and growing more quickly, in terms of both their physical size and their population. Despite the differences, there are similarities in how those with more income have greater proximity to good-quality green space.

In noting these disparities in access to natural spaces, it's worth mentioning that those least able to make the most of nature are among those who need it the most. People of lower socioeconomic status tend to reap greater benefit from urban green space than more privileged groups, especially in terms of reducing stress and improving mental health, thereby dealing a doubly positive impact on the overall health of society. This effect has also been found in relation to young and elderly people. There is also a gender dimension, with studies from Sweden suggesting that while women seem to attach more value to green areas than men, they sometimes feel less safe there, which prevents them from using them.

On top of those disparities, communities with a high proportion of immigrants and ethnic minorities are also found to have less access to high-quality green and blue spaces, with the poor quality of their local environment having a considerable impact on their health and well-being. One striking finding from a study by Friends of the Earth in England is that people of Black, Asian and minority-ethnic origin are more than twice as likely to live in areas that are most lacking in green areas compared with white people.[59] Almost 40 per cent of those from racial-minority backgrounds live in England's most green space-deprived neighbourhoods, compared with 14 per cent of white people. That 14 per cent will, in turn, be mainly those on low incomes. And in England, Black people are nearly four times as likely as white people to have no access to outdoor space at home. Even comparing people of similar age, social grade and living situation, those of Black ethnicity are 2.4 times less likely than those of white ethnicity to have a private garden.

Natural England, the country's official government nature organization, which, since 2019, it has been my privilege to lead as its chair, found in that year that 69 per cent of white people reported visiting natural spaces at least once a week, compared with 41 per cent of Black people and 38 per cent of people from an Asian ethnic background.[60] Moreover, 74 per cent of people from the highest socioeconomic groups reported visiting natural spaces at least once a week, compared with 53 per cent of people from the lowest socioeconomic groups.

Another study, by the Ramblers Association, a campaign group that promotes access to the outdoors in Britain, found that people from ethnic-minority backgrounds are less likely to live within a five-minute walk of a green space than white people (39 per cent compared with 58 per cent). They were also less likely to report that there are good walking routes where they live (38 per cent compared with 52 per cent) and less likely to report a variety of different green spaces within walking distance of where they live (46 per cent compared with 58 per cent).[61] In a similar vein, another report found that in areas where more than 40 per cent of residents are Black or minority ethnic, there is eleven times less green space than in areas where residents are largely white, and the former are likely to be of poorer quality.[62]

Rhiane Fatinikun has spent a lot of time thinking about the racial dimensions of access to green areas. Not only has she been thinking about them, she's also started a highly successful initiative to do something positive to connect people of colour with nature. It's called Black Girls Hike, and it was set up to challenge stereotypes and the lack of representation of non-white people in outdoor activities. It found strong support, and local groups have now been launched nationwide, running regular hikes, outdoor activity days and training events. I asked Rhiane about the origins of the initiative. 'I got the idea to start Black Girls Hike on the train going through the Peak District,' she told me. 'I was just watching people get on and off and they were all old and it was all white. And I remember thinking I want to do something new for my well-being and that the Peak District is really close to me.'

She said that she didn't drive, but could use the train, and that although that huge National Park should have been accessible, it wasn't. Rhiane explained how this was in some large part down to the fact that going out to walk in a National Park or other non-urban environment wasn't something that people of colour normally did. She told me that there were many reasons for this, but one of the main ones was 'historical perceptions and the location of where the majority of marginalized people tend to live in the UK', and the fact that Black and Brown people tended to live 'in urban places where they don't necessarily have access to green spaces and when they do have access to green spaces, it's not quality green spaces'.

The people she was seeking to engage more in outdoor activities came mostly from inner-city areas, and she found that the first reaction of many was to think that the outdoors wasn't for them. This was because people did not feel that they had the skills or the right clothing, never mind any sense of where to go. She added that it was also in part linked with the 'intersection between class and race' and with how the 'majority of Black marginalized people in the UK do tend to be in the low classes'. These were huge barriers, but ones that she confronted head-on, using social media to drum up participation.

'I started an Instagram page, and on the first walk we had 14 people come. We also had someone come who was a journalist for *The Voice* newspaper, which is the Black newspaper, and she put us on the front page and then we just started getting loads of demand from across the country. People asking if we had groups there and then getting loads of press attention, and brands wanting to work with us and it just kind of took off. I think for me it was kind of like the community saying, "Oh, we are hiking, this is what we're doing now" because there was that representation that made it accessible, and it made it something that people can aspire to.'

As the idea took off, so Rhiane began to see more clearly some of the factors that had tended to prevent Black people from enjoying natural areas. These included the extent to which non-white people interact with the wider society they live in.

'One of the things that I've noticed is a lot of the people in our community that live in London don't necessarily mix as much with the wider community as people for example in our northern groups, so they are much more hesitant about going into predominantly white spaces in the countryside. I grew up in Blackburn, which is classed as semi-rural I think, so I'm used to being closer to the outdoors and being in a very white environment. For some people I think this is a barrier because it makes them feel less safe.' With this thought in mind, it seems that if social barriers between racial communities living close to one another are higher, then so are the inequalities in relation to access to green space, which is a very important thing to know.

Whereas I expected the reason behind these people's feelings of vulnerability might stem from some form of racial abuse or attack, Rhiane gave a different reason. 'When it comes to safety, for most of the people that I've heard talk about it, it's not the fear of being attacked, it's the fear of something happening to them and knowing that someone is going to help them. So we did some sessions with some young people in London and it was something that one of the young girls said, that she always worried whether people would stop to help if she was injured in the outdoors because she looks different. They don't feel like they're part of the wider community there.'

These explanations draw attention to the situation faced by racial minorities in urban settings, but there are also major issues affecting some white people living in rural areas. These rather underline the reality of much of the rural landscape, which, although green in colour, can be difficult to enjoy for those walking or cycling due to the absence of paths and the lack of public access, and which, because of intensive farming, is also bereft of wildlife, with fewer birds, bees and butterflies even than are found in many urban areas. Maps produced by Natural England show clearly how large areas of rural England lack access to green space. While the density of people living there is lower than in the towns and cities, this aspect of environmental inequality is prevalent outside the towns as well as in them.[63]

Very often the planning of new housing developments, whether in areas that are already urban or on the edge of rural villages, has focused generally on the number of dwellings and the needs of motor vehicles, rather than on the overall quality of the environment. And where green space is planned in, it is sometimes comprised of open areas of grass or, even worse, plastic trees and bushes. Dead residential areas bereft of even living vegetation and the birds and insects that depend upon it fail to meet a profound need in people – namely, to connect with the nature that sustains us. Over time, this will have massive implications because, as fewer and fewer of us have direct experience and knowledge of nature, we become less and less likely to value what we are losing or have already lost. This progressive and ongoing disconnection from nature, and the inequalities linked with it, are very far from a passing concern.

NATURE DIVIDEND

The value of the benefits gained from access to nature is measurable not only in the health and well-being of individual citizens, but also in the economic upsides (and downsides) for society. The rise of chronic ill health, seen, for example, in the wide prevalence of anxiety, depression, cardiovascular disease and type 2 diabetes (all conditions that can in different ways be ameliorated through access to quality outdoor spaces) has added massively to the heavy burden on public health resources, sometimes in surprising ways. Take the impacts of type 2 diabetes. In 2021, this condition affected about four million people in the UK.[64] It has serious health consequences, including the possibility of strokes, amputations, heart disease, kidney failure and blindness, and is caused by poor diet, inactivity and weight gain.

The number of people diagnosed with type 2 diabetes roughly doubled in the 15 years preceding 2021, and a paper published by

Diabetes UK predicted that by 2030 the number of affected people could rise to 5.5 million, triggering a 'public health emergency' that would be not only disastrous for the people affected, but expensive too. In 2022, the National Health Service (NHS) said that in England alone it spent around ten billion pounds per year on diabetes, which is about a tenth of its entire budget. That proportion of health expenditure is set to rise as the number of people with the condition also rises.

A 2024 report by Diabetes UK confirmed a continuing and alarming upward trend in the diagnosis of type 2 diabetes among younger people. It found that in that year almost 168,000 people under 40 suffered from the condition, which is an increase of more than 47,000 – that is, over a third more – compared with 2016–17. This bucks the historical trend for the condition to be associated with older people. Now it is increasing among younger people faster than it is among the over-40s. And as with so many environment-related health conditions, there is a strong inequality dimension to this, with diagnoses disproportionately skewed towards those living in the most deprived areas and those with Black and South Asian backgrounds. Shockingly, children in the most deprived areas were found to be five times more likely to develop type 2 diabetes than those from the least deprived places.[65]

The pathways and causal connections are complex, but studies have found that a shortage of occupational and economic opportunities, inadequate access to healthcare services and information, and lack of availability of healthy food and places to exercise all lead to higher rates of diagnosis. And as that recent Diabetes UK study confirmed, in the richer countries, type 2 diabetes is more prevalent in lower-income groups, which, in turn, include the people with less access to quality green spaces.[66,67]

A drug called metformin is one of the most common treatments for this condition. As with some other pharmaceuticals, including certain antidepressants, metformin is excreted from the body unaltered, thus arriving in the wastewater stream and then in sewage works in the same form that it was taken as a medicine. It is very hard to catch this chemical with normal sewage treatment technology, and therefore much of it is then released into the environment, where it is now one of the most abundant pharmaceuticals found in effluent and in rivers. Research reveals that once it is in the environment, this drug impacts on the physiology of wild species, such as molluscs and fish. One US study into the effects of this drug on fathead minnows found that at concentrations now commonly found in rivers it caused feminization of the male fish, reduction in size, and lower fertility, thereby limiting their ability to breed.[68]

Metformin is one among a range of modern pharmaceuticals that have been prescribed in ever greater quantities during recent years and that are now causing major pollution-control headaches. While sewage pollution has recently been of great public concern in Britain, the question of what is actually in the sewage pouring into our rivers has been much less discussed. The fact is, though, that whereas the treatment of human waste to remove phosphates and other ecologically harmful pollutants presents a serious challenge, the addition of more and more micropollutants has made the challenge far greater still.[69] So significant is this problem that some in the water industry doubt the extent to which it will be possible to deal with it in the future, due to financial costs and the high amount of energy and carbon emissions required to make engineering and technical solutions viable. If we want our rivers to be clean, therefore, some other solution will be needed. Logically, this might include steps to ensure lower prevalence of type 2 diabetes, which, in turn, would require actions to deal with its causes, including the inactivity that is in part linked to lack of access to green spaces.

Antidepressants are also linked with the consequences of long-term disadvantage. Figures presented in the *British Medical Journal* in 2019 revealed how prescriptions for such drugs almost doubled in England during the previous decade, rising to 70.9 million issued in 2018, compared with 36 million in 2008.[70] The number has steadily increased year-on-year. *The Pharmaceutical Journal* said that in 2021–22, the number of antidepressant items prescribed reached 83.4 million.[71] As might be expected, the prescription of such drugs was higher in more deprived areas and among people of lower socioeconomic status. And as is the case with metformin, many antidepressants exit the human body unaltered, finding their way via sewage treatment works into the environment.

The known biological effects of these drugs on the environment include changes to the spawning behaviour of bivalve molluscs, altered behaviour among snails, reduced ability of cuttlefish to camouflage themselves, altered activity in amphipods and altered embryonic development of fish. The effects of these compounds are thus clearly diverse and potentially impacting a wide range of species. Their effects have been found to be very rapid, too.[72] For example, the drugs fluoxetine and fluvoxamine, prescribed to combat depression among other things, caused zebra mussels to spawn within minutes of contact.

During my childhood forays looking for wildlife, starlings were among the most common birds I would see. Since the 1970s, their numbers have plummeted by more than 80 per cent, for a variety of reasons, including, possibly, the effects of antidepressants. One study by researchers at the University of York investigated the effects of the antidepressant Prozac on starlings.[73] These birds, like many others that eat invertebrates, often forage for worms, flies and insect larvae at sewage treatment plants, where the latter pick up various pharmaceuticals, which are in turn consumed by the birds.

Dr Kathryn Arnold from the Department of Environment and Geography at the University of York measured the level of Prozac present in earthworms living in sewage treatment works and then fed captive worms a diet that included Prozac to the point where the same level of contamination was reached, before feeding them to captive starlings. Arnold said, 'The major findings were that the starlings lost their appetite and libido. Compared with the control birds that hadn't had any Prozac, they ate much less and snacked throughout the day.' She went on to add that 'research suggests that Prozac in the water system could lead to a decline in starlings breeding and if the birds are taking in fewer calories they are less likely to survive the winter, thus leading to a reduction in starling numbers.'[74]

While the full range of effects arising from pharmaceutical pollution is still not fully understood, the range of chemicals being released is becoming more widespread as the rate of prescription grows, driven in some large part by social disadvantage. The Natural History Museum estimates that some 40 per cent of the world's rivers could contain harmful levels of drugs that can have a significant impact on the health of organisms and ecosystems, causing behavioural change, hormone disruption and toxicity.[75] Nearly 2,000 different kinds of active pharmaceutical ingredients are used in human and veterinary medicine, and more and more of these are posing some threat to the natural world. The fact that more of these drugs are being prescribed to meet the rising epidemic of chronic illness is thus creating environmental damage, as well as economic cost – not only in terms of the cost of purchasing these drugs in the first place, but also in terms of the costs involved in getting them out of sewage water after they've been used.

The sharp rise in the prescription of drugs to treat the effects of inactivity and depression thus reveals a further layer of connection between disadvantage, access to green space and well-being. The first layer relates to the extent to which green spaces protect and maintain our health; the second details the positive impact exposure to natural areas has on creating a mandate for action to protect and restore the environment; the third links the benefits of better access to green areas to lower economic burdens on society; and the fourth concerns pharmaceutical pollution arising from medication use among people less likely to have easy access to green areas. On top of all that is a fifth layer, regarding the inequalities evident in how we must adapt to the already unavoidable consequences of global heating.

THE GREEN SHIELDS

Britain is known around the world for its cool, damp climate, so the extreme heat and drought experienced across the country during the summer of 2022 was indicative of the profound shifts that are under way. Ponds dried to pans of hard cracked mud. Grasses and herbs withered to brittle yellow tinder. River flow plummeted, leading to widespread fish deaths, and trees shed their leaves in summer. Many species suffered from a lack of food and water, and red weather warnings were issued, denoting not only likely damage to infrastructure such as railways, but also the risk of death.

That hot summer also revealed the importance of green spaces to people's ability to cope with the heat. In July 2022, for the first time ever in the British Isles, the temperature went above 40 degrees centigrade. Brutal and extreme, it was, according to meteorologists, an event that would have been virtually impossible in the absence of human-induced climate warming. As we saw in the last chapter, this kind of heat, especially when it lasts for some days, can lead to serious health risks. These include heat exhaustion and heatstroke, with prolonged high temperatures also contributing to deaths from heart attacks, strokes and other forms of cardiovascular disease.

Recent times have seen significantly increased mortality resulting from hot weather, with more than 60,000 premature deaths across Europe due to that heatwave in 2022.[76] In the UK during the hottest of five hot periods that occurred between June and August 2022, it was in mid-July that records were smashed and excess deaths went above ten per cent more than the five-year average (adding up to over 2,000 more deaths than would have been expected had the patterns of previous years been followed).[77]

When extreme heat does strike, one of the things that can diminish its impact is green space, and especially the presence of mature trees. Plants bring water up from the ground and evaporate it through their leaves into the air, creating a natural air-conditioning effect, while at the same time trees cast shade, which can help prevent the ground, and then the air, from warming up so much. Surfaces covered with concrete and buildings, by contrast, absorb and store heat. Then, like a radiator, they release it, causing more extreme conditions than would otherwise have occurred. These different properties of built areas versus green spaces are reflected in often quite large temperature differences between cities and towns and their rural surroundings.

New York City, for example, is generally about four degrees centigrade warmer than the non-built-up areas outside, and while that might not seem much, it can make a huge difference when temperatures climb like they did in England during the summer of 2022. Sometimes the temperature difference can be considerably more than that – London can be up to ten degrees centigrade hotter than its rural surroundings during heatwaves.[78] People on lower incomes and those disadvantaged in other ways are not only less likely to live close to green spaces, they are also less likely to have a garden. They are more likely to live in rented accommodation and have limited means to adapt to such extreme heat or exert any control over the situation. Considering the importance of trees and other vegetation in enabling people to avoid the worst impacts of high temperatures, we shouldn't underestimate the significance of the 2023 research by Friends of the Earth revealing how areas with the highest levels of social deprivation have far fewer trees than the wealthiest neighbourhoods.[79]

Similar patterns have been detected in the USA, where the disparities between green and less green neighbourhoods have been linked with deliberate spatial planning policies. For example, in some of the areas in which historical 'redlining' policies prevented people of colour from buying homes (and this is also linked to such people's exposure to higher levels of toxic pollution), the extent of tree canopy was a major factor. One recent study found that in 37 US cities, areas previously subject to such segregation have 23 per cent less tree canopy than elsewhere, with a different study finding that 94 per cent of such districts in 108 US cities had elevated temperatures compared with the rest of the city within which they sat.[80,81]

There is also evidence of a connection between the species diversity in green spaces and the cooling effect they provide, making those with a wider variety of species better able to reduce the heat-island effect while elevating the biodiversity that is so beneficial for psychological health, as we have already seen. For example, researchers in China looking at the character of green spaces found connections between tree-species richness, canopy coverage and temperature amelioration, with results indicating that tree-species diversity can explain up to 59 per cent of the cooling variation. Other research demonstrates how increased vegetation complexity reduces surface temperatures, with flower meadows, hedgerows and shrubs exhibiting greater cooling ability than standard lawns.[82]

Across the world, destroyed or degraded green areas leave people suffering from the effects of high temperatures, but exposed in other ways too. The clearance of forests from hillsides increases the risk of landslides, while the loss of coastal mangroves can leave communities naked in the face of the extreme weather driven by climate change. As we have seen, in the wake of Hurricane Katrina, the impact was made much worse through the degradation of coastal wetlands. Those who tend to suffer first and most in such situations are the least well-off, and, in that instance, they were predominantly Black people.

Research from around the world reinforces this connection between ecological degradation and vulnerability to the impacts of climate change. It is the least well-off who tend to feel the effects first and hardest. For example, in a 2023 study looking at the protective benefits provided by Colombia's natural ecosystems, the United Nations Environment Programme World Conservation Monitoring Centre (UNEP-WCMC) found that the progressive loss of wetlands, mangroves and coral reefs will increase the vulnerability of low-income communities as the protection that these systems provide is diminished.[83]

Inequalities in our environmental experiences and our access to high-quality green spaces and the uneven way in which the impacts of climate change fall across different parts of society are in turn linked to the hugely varied ways in which different groups of people cause environmental damage in the first place. We'll turn to that subject next.

7

Global Stand-off

When I first engaged with the global environmental political process during the early 1990s, I was a young campaigner focused very much on environmental issues – deforestation, the extinction of species and climate change among them. I had yet to realize, however, that making progress on these questions required far more than just good environmental science or brilliant ideas for new green policies.

There were various moments when this realization became an increasingly vivid insight. One was in 1990 when, while working with BirdLife International, I went to the north-east of Brazil with scientists from that country in search of what was to prove to be the world's rarest bird: the Spix's macaw.[1] This blue parrot had always been rare, but by then it was thought to possibly be extinct. Our team of five searched widely across the parched interior of this poverty-stricken region, travelling for weeks across country along dusty tracks in an old Land Rover and a Toyota, to scour remote areas of remaining natural habitat that even by then had been fragmented and degraded by vast tracts of soybeans destined to feed pigs and chickens in distant factory farms. Following the removal of the forests to make way for the huge fields, the fragile tropical soils had been washed away by torrential seasonal rain.

The massive agribusinesses that were fuelling Brazil's export-led economy were expanding across landscapes populated by desperately poor people. Many of them had few rights or legal entitlements and thus were effectively swept aside by the massive farms producing commodity crops destined for overseas markets. I came to know one such family who lived in a tiny thatched house next to a remote creek where we finally found one of the birds we were looking for. Their home was set in a dramatic rugged landscape, with cacti, patches of thorny woodland and scrubby grasslands. Big trees grew by the side of the creek. We paid them a little to cook us meals and they slaughtered one of their goats for us to eat. During the ten days or so that we were there we consumed the whole animal, starting with the choice cuts and finishing with a stew that contained its feet.

That bird was so rare not only because of the gradual loss of its habitat to agriculture but also because collectors of rare parrots were willing to pay tens of thousands of dollars for each one via a sophisticated network of trappers and traffickers who got the birds to international buyers. Big money was being made from the exploitation of Brazil's soils and wildlife, but it was striking how few of the benefits went to the local people. The natural environment where they lived was being liquidated and exported for lucrative financial reward, while they scratched a living growing corn and tending a few animals. They had no power, no influence and no voice.

My head was crowded with many questions. How could the plight of the many poor farmers who lived at the edge of these dwindling wild lands be improved at the same time as efforts were being made to recover endangered and depleted wildlife? Could they enjoy more wealth without causing environmental damage? What could be done to share the economic benefits derived from exploiting Brazil's natural resources more fairly?

Many practical challenges follow in the wake of reflections like that, but in the end I came to see that it was matters of fairness, justice and equality that were at the heart of the matter. And entwined with that is how people tend to demand more from nature the wealthier they become, be they commodity companies or collectors of rare birds. Take the emissions of greenhouse gases, which come not only from fossil fuels, but also from deforestation. The contributions different groups in society have made to this most fundamental driver of environmental change are far from equal.

One measure of the disparities in play is the calculation that the richest one per cent of people are responsible for more than twice as much greenhouse gas pollution as the poorest 50 per cent of humankind combined.[2] Even if we look at an average US citizen, the way they live leads to emissions that are over five hundred times greater than those caused by an average citizen of the small East African country of Burundi.[3] This massive difference exists, of course, because the average American has use of a vehicle, lives in a home with heating, cooling and electricity, consumes a diet rich in meat and dairy (in turn partly sustained by soybeans produced on land recently deforested to meet global demand), and has access to a vast range of consumer goods and flights on planes. By contrast, most people living in East Africa (and many other parts of the world where average incomes are much lower) conduct their lives with far less, if any, of that.

The simple fact is that as incomes rise, people consume more energy and resources, which in turn causes more environmental pressures. A study from Oxfam concluded that the consumption-based emissions (that is, those arising from what people use and how they live, rather than from the fossil fuels their country exports) of the richest ten per cent of the world's population in 2015 accounted for nearly half the global total, with other research putting the figure at between 36 and 47 per cent.[4,5]

These disparities are evident not only in the here and now, but also over time. James Hansen, director of the NASA Goddard Institute for Space Studies and Adjunct Professor at the Columbia University Earth Institute, concluded that more than three quarters of the climate-changing emissions released into the Earth's atmosphere between 1751 and 2006 were caused by the fifth of the world's population who lived in Europe, North America, Australia and Japan.[6]

In 2015, Oxfam published research that conservatively estimated that the average emissions of someone in the poorest ten per cent of the global population are 60 times less than those of someone in the richest ten per cent. A separate analysis by Lucas Chancel, published in *Nature* in 2022, estimated that while the bottom 20 per cent of the world's population release on average about 0.4 tonnes of carbon each per year, the richest one per cent release 101 tonnes each, with the super elite in the top 0.01 per cent contributing 2,332 tonnes each. In percentage terms, the top one per cent of humankind are responsible for nearly 16 per cent of emissions, while the bottom half is responsible for just over a tenth (11.5 per cent).

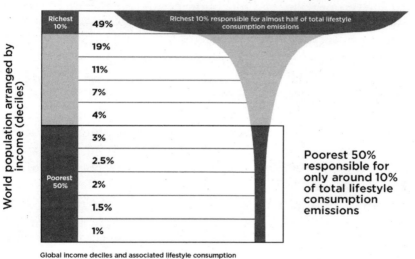

Figure 9

Increased income for those living in poverty brings huge benefits, but if the tendency is for such an increase to exacerbate environmental pressures, then some fundamental questions must be posed. To what extent can our present approach to economic development be sustained? How can wealth creation be reconciled with environmental goals? Should the approach be to keep the poor in poverty, or should the better-off reduce their impact, creating space for those with less to catch up? The alternative to facing these conundrums (and, by and large, the approach we are currently taking) is to ignore the questions and proceed as if increasing consumption and environmental impact in line with wealth creation isn't an issue at all.

Susan Smith, the geographer and academic whom we met earlier, has delved deeply into social inequalities during a long research career, and she sees fundamental linkages between these social and ecological questions. 'High levels of inequality go hand in hand with concentrations of carbon. It's startling how similar the figures are. You could almost swap a graph showing the concentration of income and wealth into the hands of the top ten and one per cent, and [one showing] the proportion of carbon emissions accounted for by the top ten and one per cent.' Not only does she see the phenomena of growing inequality and rising concentrations of carbon dioxide as correlated, but she also argues that they are causally linked. 'Governance structures that support the very high levels of economic inequality that we're witnessing today are also the regimes that are producing an environmental disaster. It seems to me that if we don't do something about levels of economic inequality, there will be another turning point, and it will be caused by a catastrophe, and that catastrophe will have to do with climate.'

Carbon emissions are a very clear example of disparities in environmental impacts, but they are not the only aspect of this issue. As wealth grows, so does demand for land, water and materials. One manifestation of how our global environmental footprint is growing is seen in the rocketing demand for the resources needed to manufacture all the varied trappings of our modern world. These include minerals, ores, biomass and fossil fuels, the latter harnessed not only as an energy source, but also as feedstock for paints, plastics and other components of modern consumer goods.

Since 1970, the consumption of these materials (not including food and water) has quadrupled, reaching a total of about 100 billion tonnes in 2020.[7] According to the United Nations, 'Without concerted political action, it is projected to grow to 190 billion metric tons by 2060.'[8] Considering the scale of damage resulting from present levels of resource extraction, processing, use and disposal, maintaining present (never mind future) demand while achieving our ecological goals requires a rethink about how growing needs are to be met. As is the case with carbon emissions, there is, unsurprisingly, a striking equity dimension that lies behind the headline trend: again, the wealthy use up much more per person than those on lower incomes.

MATERIAL ISSUES

Across the world, average material consumption between 2000 and 2017 rose from 8.8 to 12.2 tonnes per person per year, with the quantity of resources consumed varying hugely between the richest and poorest.[9] In 2000, people living in low-income countries (for example, much of sub-Saharan Africa) used on average 1.4 tonnes each, rising to 2.0 tonnes per person in 2017. By contrast, citizens of high-income countries (such as in Europe and North America) used 25.6 tonnes each in 2000, rising to 26.3 tonnes per person in 2017. These figures reveal how an average citizen living in, for example, Mozambique is using just one thirteenth of the resources of an average citizen in a European country.

On top of the huge disparity in material consumption between rich and poor, more developed countries are heavily reliant on imports to sustain their expanding demand for resources. The UN estimates that on a per capita basis, high-income countries rely on 9.8 metric tons of imported primary materials per person per year, extracted elsewhere in the world and sometimes involving serious environmental impacts, which, in turn, as we have seen, tend to fall hardest on the poorest and most vulnerable.

Another critical resource that underpins societies globally is water, and that too is consumed very differently depending on wealth. Fresh water is needed in the production of all kinds of manufactured goods, as well as being vital in the energy and food sectors. The more materials, food and energy that are consumed, the more water we need to enable this to happen. The more water that is extracted, the greater the environmental pressures that can result. Many water-dependent ecosystems are damaged through over-abstraction, as stream flows drop and wetlands dry out.

There is also an important dimension linked with water quality, and the extent to which rivers, wetlands and the sea become more polluted in parallel with rising human consumption. For instance, growing demand for paper packaging means that more and bigger pulp and paper mills are being built to process raw wood, which in turn requires larger quantities of water, which in some places is returned to the environment in a highly polluted state.[10] The same can be said for the garment industry that supplies billions of items of apparel to the burgeoning cheap-clothes market.

When it comes to what we eat, not all food is equal. As wealth goes up, so does our consumption of meat and dairy-based foods, which in turn require not only more energy to produce, but also more water. This tendency towards more livestock consumption occurs at a relatively low level of wealth, but it is nonetheless part of a trend which sees an increase in environmental impact derived from the choices that come with rising income. Animal-rich diets also require more land to sustain them than diets based more on plants.

Increased per capita demand for land is linked not only with diets, but also with how people spend their spare time, with some activities requiring quite a lot of it. Take horse riding, and the joy and pleasure that come with a pursuit based on these large animals being kept for recreation. It's a lovely thing to do, and from time to time I've had the pleasure of participating myself, but it does take a lot of space to keep the animals and to grow enough food for them. Indeed, one estimate is that about half a million hectares of the UK are devoted to keeping horses.[11]

Golf is another activity that requires a lot of land and, like horse riding, is a pastime more available to the better-off. In England, about 150,000 hectares are devoted to this sport.[12] The city of Birmingham covers about 26,000 hectares, so golf courses take up five times that space, and the land needed to sustain horses about twenty times as much. If everyone engaged in such pastimes, we'd soon be short of land. Is it best to devote space to such activities, or is it better to use the land to grow more crops, thereby contributing to food security, or should it be used to create more wild areas to help reverse the decline of nature and catch carbon?

In the end, the tensions between our rising demand for resources and the resulting escalating environmental impact come down to there being limited 'environmental space' to accommodate expanding consumption. Environmental space includes the resources extracted to manufacture consumer goods and the places where we can dispose of our waste, such as carbon dioxide into the atmosphere and nutrients into the ocean. It also includes the physical space taken up by fields, reservoirs, pastures, plantations and open spaces needed to meet our needs for food, wood, water and recreation. It is a useful term when considering how land comes under increasing pressure as the population and wealth both grow, with one (wealth) multiplying the impact of the other (the population).

With limited environmental space, it is no surprise that our collective impact has increased as the economy has grown and people have on average become wealthier. This is the basic driver that lies behind the Great Acceleration, which generated the onset of the Anthropocene epoch. I realize that these observations are challenging. I present them not in support of an ideological position or to convey blame, but simply to highlight that we are exceeding the physical capacities of our planet against a backdrop of very unequal consumption, which is a material factor limiting our ability to act.

One way of looking at the scale of the mismatch between what we are using and what the planet can indefinitely sustain is to calculate how many Earths we would need to meet our demands. Numbers based on 2018 data from the Global Footprint Network concluded that the present total human demand for environmental space would require 1.8 Earths to maintain that demand indefinitely.[13] If the global population lived like the British do, then the amount of environmental space needed would add up to 2.6 Earths, rising to 5.1 Earths if everyone lived like Americans.

The inhabitants of many poorer countries, however, still live below the one-planet threshold, including those of India. According to the Global Footprint Network, if everyone lived the life of an average Indian, 0.8 Earths would be needed to meet humanity's total needs. Countries that are poorer still, such as some of those in sub-Saharan Africa, remain well within planetary limits, with their environmental footprint adding up to 0.5 Earths, if we all lived as their people do.

Looking at these kinds of analyses, it seems to me that one unavoidable and rather simple conclusion is that if we are to avert an environmental catastrophe, humankind can't collectively consume as the wealthy presently do. By 'wealthy', I don't just mean the super-rich, I'm also referring to those with middle-class western lifestyles. The level of consumption required to sustain, never mind expand, those lifestyles cannot be met on an ongoing basis.

Let me stress again that this is not an ideological statement, but a fact of life in a world with an insufficient amount of atmosphere, ocean, land and ecosystems to keep up with more and more people living like most of us do in Europe, North America and other developed countries. This stark reality has a very challenging political face, for not only is there not enough environmental space to meet the growing demands of those who are better off, there is also not enough for those who presently have very little and who *need* to increase consumption. This in turn forces us to bring questions of fairness and justice to centre stage.

STALEMATE

Unsurprisingly, those living in poverty and those lucky enough to be quite well-off have very different takes on environmental questions. I have repeatedly seen how this clash of contexts plays out in ways that have weakened the substance of global agreements, as well as the appetite of countries to implement them even when they are adopted. The disagreements have been founded on some big questions: Who should take what action by when? What should be paid by whom to compensate for environmental damage caused by present and past consumption? And who has the moral responsibility to show leadership in reducing environmental damage?

The difficulties in reaching consensus on these crucial questions in large part explain the situation I described earlier, whereby global accords have not been delivered, despite the rising body of science that warns of an increasingly dire picture. The dynamic at play has been evident since before the Rio summit in 1992, visible in the different attitude of the richer countries and their counterparts from what became known as the Global South (that is, the developing nations of Africa, Latin America and Asia).

Sitting through weeks of negotiations in Geneva and then at the Rio summit itself, I saw how the stand-off between developed and developing countries was the main barrier to progress. Representatives of countries from the Global North (Europe, North America, Japan and other developed countries) tended to speak about environmental issues and the need for action on deforestation, wildlife, climate change and pollution. This was not least because of the rising clamour for action back home, with ministers under pressure to press for ambitious agreements.

Meanwhile, the developing countries, led by, among others, Malaysia, Brazil and India, focused more on poverty and development. Their narrative focused largely on the historical role of richer nations, which, they said, had been mainly responsible for causing the problems that they were now urging action on, including the climate change that came in the wake of the Industrial Revolution. They pointed to how those who said they wanted action to cut pollution were the ones who'd put most of it into the atmosphere in the first place, and who had also cut down their own forests centuries ago. They highlighted the massive inequalities between the richer and poorer countries and stressed how these needed to be recognized in the agreements being negotiated.

That recognition would need to be seen, they said, in the rich countries paying for less well-off nations to make transitions towards greener outcomes, in the transfer of clean technologies from rich to poorer countries, and in the Global North showing leadership by cutting its own emissions and other environmental impacts first.

After all, they said, those countries had largely caused the problems through the industrialization they had embarked upon nearly two centuries before, so they should show their commitment not only by acting first but also by providing large financial resources to both compensate for the damage already done and to assist in adapting to the climate changes already locked in.

The developing nations also pointed out that they were still emerging from poverty and needed to focus on economic growth, not environmental goals. It was economic growth that had made the industrialized nations rich, they said, and now it was their turn to catch up. The development of high consumption to drive economic growth had, after all, enabled Britain, the USA and Japan to become powerful global players, and now the developing nations wanted to have their chance to bring comfort and prosperity to their people too.

Some of the most vocal developing countries were already very unequal themselves, with Brazil and India massively divided in terms of the wealth enjoyed (or not) by their own populations. The context described by the negotiators, however, was broadly correct, and although economic disparities between countries have tended to diminish since then (while increasing within countries), it still broadly is.

For example, the 2015 Oxfam analysis I mentioned earlier revealed how the majority of the world's richest ten per cent of high emitters lived in OECD countries,[14] with around a third of them in the USA. More than two decades after the Rio summit, China had caught up somewhat, although a large share of its emissions were down to the production of goods consumed in the rich countries. When it comes to what is now the world's most populous country, India, the per capita emissions of the richest ten per cent of its citizens are just one quarter of those of the poorest 50 per cent of US citizens, while the poorest 50 per cent of Indians have a carbon footprint that is just one twentieth of that of the poorest 50 per cent in the USA. So, while wealth inequalities between countries have diminished, massive inequalities in environmental impact still remain between them, with the rich in the poor countries way behind even the poorest of the rich countries. So even most of the better-off people in developing countries have emissions that are below those caused by the poor in rich countries.

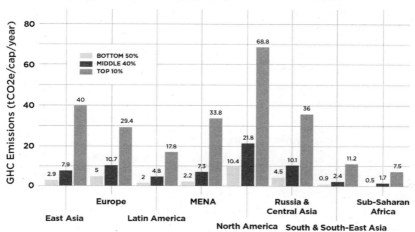

Figure 10

Against the backdrop of massively unequal contributions to environmental challenges, there was a sense among the developing countries that some would rather the poor stayed poor in order to protect the environment. Unsurprisingly, they regarded this as an unjust and untenable context for talks. They also reacted strongly against the idea of population growth being the main problem which had for decades been one of the principal conclusions of many western environmentalists, including some very prominent ones.

Those environmentalists claimed that the population explosion after the Second World War was the biggest driver of ecological damage, and to an extent they were not wrong, though there were several very important qualifications and caveats to this conclusion which sometimes became obscured behind the focus on large family sizes. This was generally interpreted as being more of an issue for poorer countries, rather than developed nations. During the period between the 1960s and the 1990s, when population growth was very strong, it was indeed in the developing countries that populations grew fastest, but this focus led delegates from these countries to worry that it was a way of blaming them for the global ecological crisis.

The developing countries retaliated by pointing out how it was not population per se that was the main driver, but the per capita impacts arising from individual consumption. This was more important than the number of people, they maintained: what people did was the real issue, not how many of them there were. With an average US citizen emitting as much as five hundred times more carbon than a poor person in Central Africa, they had a point. There were also voices highlighting how it was not only a matter of more children being born, but also one of rising longevity, particularly in the wealthy western countries, where the baby-boomer generation saw not only ever-rising living standards, but also, at that point, ever-longer lives.

In North America, life expectancy rose from 68 years in 1950–55 to 78 during 2005–10, which is about the level it remains at today. By contrast, the life expectancy of an African person during the early 1950s was 37, rising to about 56 in 2005–10.[15] In 2021, the disparity between the country with the highest life expectancy at birth and the country with the lowest stood at 33.4 years. That disparity is underlined by 2021 figures revealing how life expectancy at birth in that year was close to 85 years or above in Australia, Japan and a few other developed countries, while in the Central African Republic, Chad, Lesotho and Nigeria, it was below 54 years.[16]

This disparity is very important because people not dying is obviously a significant component of overall population growth. And the rich with their longer lives have a far greater per capita consumption than an average African baby, who will not only use fewer resources but will also expect to have a shorter life than a western infant. For these reasons, the focus on population growth as the problem, which placed the onus on developing countries, became a red rag for them in negotiations.

Alongside disparities in consumption patterns, some people are far more exposed than others to the effects of the pollution caused mostly by the richer members of global society. In 2023, most of the countries with the fastest-growing populations were in sub-Saharan Africa, and although between them they are responsible for a tiny fraction of emissions, at the same time they are suffering from some of the most serious impacts of global heating.[17]

The population and consumption debates, and the differences that emerged between countries as the negotiations intensified towards the Rio summit, were therefore a critically important backdrop to the environmental agreements reached there, and shaped the likelihood of their subsequent implementation. The mood of the talks was also influenced by calls to curb the kind of high consumption and energy intensive economic development that had already made the developed countries rich. In the face of the environmental crisis, any proposal to end that kind of economic growth was seen as a further attack on developing nations, and an attempt to keep their populations destitute. Whether or not this was a correct interpretation, it nonetheless created yet another barrier to progress.

Writer Geoffrey Lean, who covered the Rio summit and the negotiations leading up to it for the *Observer* newspaper, recalled some of the big ideas in play, and how they became manifest in the politics of this critical global moment. He described how two groups of western advocates came to the discussion with very different takes on what was needed. One promoted the idea of sustainable development – that is, the notion that came from the Brundtland Commission's report, *Our Common Future*, mentioned earlier. 'This was the route to development that was utterly and fundamentally connected with justice, world development and poverty,' he explained. In this school of thought, environmental, social and economic plans needed to be joined up, pursued as a single endeavour and not traded off against one another. This idea had origins going back to the 1972 Stockholm Summit on the environment and development, where the Indian prime minister Indira Gandhi made her famous speech that included the question, 'Are not poverty and need the greatest polluters?' Later requoted as 'Poverty is the worst form of pollution', it was a simple but powerful thought that would shape the discussions in Rio, linking development to environmental goals.

By contrast, the other faction, according to Lean, 'was sceptical of development and poor people developing because that would ruin the environment.' This was a significant divide, and at the time hugely influential publications shaping the discussion were drawing support towards the latter view. One of these was a book called *The Limits to Growth*. It had been published by the Club of Rome in 1972, but 20 years later it was still influential, setting out how population growth and economic growth were taking the world towards ecological disaster. It agreed that everyone's needs should be met sustainably, but it was its anti-growth messages which gained more prominence. Another of these texts was *A Blueprint for Survival*, also published in 1972, as a special edition of *The Ecologist* magazine, and which recommended that economic development should be abandoned in favour of people living in small, decentralized and largely deindustrialized communities.

These publications attracted a lot of support and were powerful drivers of what emerged as prominent anti-development and anti-population growth narratives that shaped the overall approach towards global environmental challenges in the run-up to the Rio summit. Lean told me, 'The "no growth" view got huge publicity at the time, and thus shaped many people's first impressions of environmentalism.' And it was those arguments that prevailed, winning in the public mind to the point where 'environmentalism and growth were seen to be opposed to each other from very early on'.

While many environmentalists (including Friends of the Earth) did focus on equity and fairness, rather than on growth or not, the idea that development was the problem became stuck fast in the discussion. Lean puts it like this: 'I'd say the lack of recognition of equality issues in much of the environment movement [...] has actually hugely hindered environmental progress, and you still get it everywhere, you know, growth versus environment.'

The idea was thereby reinforced in the minds of many of the representatives of developing countries – and, via the media, in the minds of many of their citizens too – that addressing global environmental challenges was about stopping their development, blaming the poor and letting the rich off the hook for their historical contributions. It was not what environmental campaigners necessarily intended to say, but it was nonetheless what was widely heard, and, unsurprisingly, viewed as an unacceptable starting point from which to make the transitions needed to address environmental challenges. It is this impasse that has dogged progress ever since, despite various commitments in different treaties to equity, justice and sustainable development. In the end, the UN came up with the notion of 'common but differentiated responsibilities' in an effort to contain the stand-off and enable progress to take place.

During my time as vice chair of Friends of the Earth International I experienced many moments in which science met questions of fairness, and generally found that the latter trumped the former. For example, at the climate summit in Bali in 2007, Friends of the Earth campaigners from countries across the globe called for a limit of two degrees of global warming to be adopted as the official goal enshrined in a UN agreement. However, my colleagues from Africa, Asia and Latin America said they would be unable to support such a goal, even if it was backed by the latest science, if the route to reaching it was not founded on principles of justice.

This would need to include, they said, recognition of historical responsibility, which would mean that the countries that had mainly caused the problem (and that were still doing so) would have to shoulder the main share of action. It would need to embrace compensation for the damage already being caused by historical pollution, and be based on the transfer of the money needed to achieve sustainable development. I asked them if an agreement that would limit warming to two degrees (and thus avoid global disaster) could be acceptable without all that. 'No,' they said, confirming that as far as they were concerned, ecological disaster would be the consequence of injustice. From what I have seen over many years, that internal Friends of the Earth International policy discussion reflects a wider political reality that remains firmly embedded today.

DEEP ROOTS

For many environmentalists, the massive disparities between the footprints of different countries have deep historical roots, linked not only with industrialization, but also with colonialism and, to an extent, racism, given how members of Indigenous societies were in many places enslaved and nearly, or actually, wiped out. That process of cultural domination was in large part motivated by the desire to appropriate the resources, including the ancestral lands, rivers, seas and mountain ranges, of Indigenous peoples.

During the course of her many years of influential work as an environmental scientist and activist ecologist, Indian campaigner Vandana Shiva has reached some important conclusions. I have known Vandana for years and have had the privilege of working with her, including on a visit to New Delhi, when we shared the platform in discussions about sustainable farming. She is articulate to a degree that I have rarely witnessed and devastatingly clear in her analysis, including on the paralyzingly complex links between our modern ecological crisis and the historical events and trends in which it has its origins. These include the link between industrialization and colonialism and today's disparities between wealthy and developing countries, and the way in which private ownership replaced the management of nature by local communities.

Her views were partly shaped by the struggle of the Chipko communities living in the valleys of the Himalayas of northern India in the state of Uttaranchal (now Uttarakhand). It was here during the 1970s that a protest movement formed to recover community control of forests from policies that were a hangover from the British colonial era, with areas of rich woodland having been wrested from local control and auctioned to contractors for logging. These forestry practices had led to the destruction of the ecosystem, not only depriving local people of a source of livelihood, but also causing severe landslides and downstream flooding.

Vandana told me about her visits to the region, where she got to know the women who organized the Chipko campaigns. Becoming famous for embracing trees to protect them from the chainsaws, they were the original tree huggers, pioneering a form of protest that not only inspired her, but which travelled around the world. Vandana described how women in the remote villages had no formal education or money, but they had lots of knowledge and a very prosperous economy. 'The forest was sacred for them, but they were also protecting the soil and forests as the economic base of survival. For them the issues of justice and the issues of sustainability were never separate aspects.' Vandana believes that a large part of the challenge comes down to how Indigenous worldviews have been progressively replaced, including in relation to community control of resources. 'Land was never private property in India, until British colonialism. We had very clear use rights, one about property rights and one about use of the property rights.'

She described how the Chipko movement was just one example among thousands across the world of Indigenous and other local communities being dispossessed of the basic right to access their traditional resources and lands: 'They had the food they needed, they had the water they needed, they had the community they needed, they had the sovereignty and the local democracy that they needed. And now they have none of that. They don't have livelihoods. They don't have access to the basic things of life.' She told me how this removal of control related to some of the environmental injustices I have touched on already. 'The images are always shown of the poorest people living around the most polluted places, [but] they didn't create that pollution, whether it's in America, with the cancer alley, or the toxic waste that is dumped on Black communities.'

For her these observations raise some fundamental issues as to how we might approach environmental challenges. 'I've realized over time that the part that's missing in most ecological movements is about natural resources, who do they belong to? How are they used? How are they distributed? How are they shared? What happens when they are appropriated?' These are the questions that Vandana sees as being at the heart of the global deadlock, and yet in discussions in global political processes and the debates that surround them, they are generally not the questions on the table, as negotiators trade words about targets, complex technical rules, deadlines and reporting requirements. So big is this underlying problem that thus far it has proved impossible to resolve.

GOING GREEN

This fundamental, difficult and rather uncomfortable context is a major reason why the more immediate and technical aspects of environmental challenges, rather than those linked with deeper matters of fairness, justice and equality, are generally where attention is focused. This in turn tends towards solutions that are similarly detached from the deeper context. Take the response to climate change, and how some of what are now presented as greener choices can replicate the very inequalities that are at the root of the problem. A case in point is presented by some aspects of the rise of electric cars. The manufacture of such vehicles has quite correctly expanded following rising concerns about the impact of climate-changing emissions from fossil fuels, including those coming from the billion and a half or so internal combustion engines that drive the world's car, truck and van fleet.

The decarbonization of transport is a vital step the world must take, and using zero-carbon power sources, such as solar and wind generation, to recharge batteries is one critical pathway required to make the transition. This is necessary and urgent, but some of what is required can be accompanied by a perpetuation of existing inequalities in the consumption of materials, and by destructive impacts arising from their extraction and processing. Instead of diesel and petrol, inequalities in the consumption of minerals including lithium, cobalt, copper, nickel and aluminium to make batteries, electric motors and vehicle bodies are set to occur.[18] Mines for the extraction of these and other vital elements can cause deforestation, and local populations can be exposed to toxic substances through air and groundwater contamination.[19] Once ores are extracted, significant amounts of energy and water are consumed to refine and process the minerals into a useable form. In order to meet rapidly expanding demand, hundreds of new mines will be needed, with all the pollution and ecological damage that can come with such operations.[20]

Many of these mines will be on land, and increasingly there will be growing pressure to obtain minerals from the bottom of the ocean, including via the exploitation of billions of potato-sized nodules of material scattered across seabeds worldwide which contain multiple metals, including cobalt, magnesium and nickel.[21] As with mining on land, extracting these nodules will inevitably be accompanied by environmental impacts, including the disturbance of deep ocean sediments, noise pollution and damage to habitats.

Key to fast-growing low-emissions technologies is the need to manufacture many millions of rechargeable lithium-ion batteries. Lithium is found in high concentrations in a few countries, including parts of South America (particularly Bolivia and Chile), China, Australia and the USA.[22] In common with other resource-extraction industries, those involving lithium, and also cobalt, have attracted international attention. Reports of child labour in cobalt mines in the Democratic Republic of Congo, as well as in lithium-mining projects in South America, have made headlines globally.[23] The environmental and social downsides can be minimized, but will still lead to damage that would not occur were it not for the rising demand for raw materials being driven by wealthy consumers, including those quite understandably motivated by environmental concerns. But switching from one set of resources (fossil fuels) to another (minerals, including cobalt and lithium) in order to sustain and grow mass private vehicle use perpetuates what in the end can still be unsustainable and inequitable patterns of consumption. This conundrum has been registered at the highest level.

In April 2024, the United Nations launched a new process to find ways of avoiding the risk that the transition to greener technologies that is now under way may well repeat some of the same problems it is seeking to solve. The Secretary-General's Panel on Critical Energy Transition Minerals aims to support a just transition to new technologies, ensure local people benefit from resource extraction and establish international standards and cooperation to do that.[24] At its launch, António Guterres, the UN Secretary-General, remarked that in a world increasingly powered by renewables there were many benefits for jobs, economic diversification and energy security, but he warned that these benefits would only be secured if the transition was managed properly. 'The race to net zero cannot trample over the poor . . . we must guide it towards justice,' he said.

And there is a further thought to bear in mind as we extract vast quantities of natural resources to support the transition to a net-zero future, including through the electrification of transport. Such vehicles are not carbon free, even when the batteries are charged from renewable electricity, as extracting one tonne of lithium from hard rock deposits leads to, on average, 15 tonnes of carbon dioxide going into the atmosphere.[25] Battery manufacture and making the parts for the rest of the vehicle add to total emissions too. When examined from beginning to end, the lifetime climate impact from an electric car presently works out at about half that of an average modern EU diesel or petrol car, and in some cases significantly more.[26,27]

With the climate science saying that the target we must achieve is a net-zero level of climate-heating pollution, cutting emissions by 50 per cent makes a massive difference, but it only takes us at best halfway. Technology is being refined, and progress can certainly be made in closing that gap, including through making the manufacture of vehicles and other consumer goods a circular process, whereby components are engineered for reuse, or at least recycling. In the meantime, and as those processes are being refined, there still are many good reasons to have an electric vehicle rather than a fossil fuelled one.

Recognizing the limitations of present technology and industrial processes, however, several sectors, many companies and some individuals have sought to make up the difference between net-zero aspirations and the current reality with carbon offsets. Carbon offsets are a mechanism whereby pollution in one place can be compensated for by paying for positive action in another. Offsets include investments in renewable power, energy-efficiency programmes and the protection and restoration of ecosystems, such as forests. It is a concept that can have merits, generating funding for useful activities that would otherwise not occur, but it can be fraught with complexities and difficulties. These difficulties arise when we start considering what can be counted as an offset – whether we have reliable ways of measuring the effect of what is being done, or if it is causing other environmental problems or creating issues for communities, and, not least, whether what is being done is in addition to what would have happened anyway. For example, if a forest is already protected, then counting it as a carbon gain when it is not cut down is a case of false accounting. These and other questions have rendered offsetting schemes highly controversial.

Laura Fox, my excellent research assistant for this book, told me about time she spent in the West African country of Liberia, working with the conservation group Fauna & Flora (I am a member of the council of trustees for this organization, so know their work to be of the highest quality). She assisted farmers in efforts to reduce deforestation, thereby generating benefits that could be offered for sale as carbon offsets. The main threat to the forest was land clearance caused by traditional shifting-agriculture systems. The plan was to reduce emissions by changing farming practices, so that food production took place on permanent plots. This required the training of farming communities to use the same land and to not only produce food for subsistence, but also to gain more income from tree-derived cash crops, including cocoa, coffee and palm oil, sold into the market.

Working with these communities, Laura was struck by how different their lives were to those of the people whose carbon emissions would be offset. 'They have never owned a car and very likely never would. They could count the number of times in their lives they have been in cars in the form of a taxi, and the closest paved road was over 50 kilometres away. They didn't have, and realistically never would have, a washing machine, glass windows, flushing toilets, and only the wealthiest in a village will have an electric generator with a power switch on the wall. Their homes have dirt floors, wooden shutters over their windows and an outdoor kitchen. They wash their clothes in the river. You could say they are not part of the consumer society.' The plan was nonetheless to involve these very poor people in a scheme to cut emissions coming from consumer societies.

Laura told me how the Fauna & Flora team, with their local NGO partner, SADS (Social & Agricultural Development Services), took the communities and farmers through a long process of community meetings to help them understand how there is a gas called carbon dioxide building up in our atmosphere, which is altering the Earth's climate. She found that the farmers were aware of climate change, not least through the changing patterns of rainfall impacting their crops, although this was not their primary reason for joining the project. Their motivation for entering into a carbon-reduction scheme was to receive support for their farming system, to develop a community fund for training people in the village to become nurses and teachers, and to build infrastructure.

The circumstances experienced by these farmers and by those buying carbon offsets could not be more different, raising questions as to whether it would be fairer if the polluters changed their behaviour first. At the same time as making that point, though, it is important to know that if it is done well, this kind of approach can make a positive difference, with one review into the effectiveness of carbon-offset projects finding that it was exactly this type of community development that yielded the best results.[28]

For carbon-challenged sectors with few immediate prospects of making major progress in reducing their emissions via technological solutions, offsetting has emerged as a particularly important policy. Aviation is one of them, with airlines seeking to find ways of compensating for the pollution they cause. Not all offsets are as well done as the situation in Liberia, however, and controversy has from time to time broken out.

In 2023, Delta Air Lines faced a lawsuit filed by a Californian customer who bought a flight ticket and challenged company claims about a one-billion-dollar investment announced in 2020 to render its business carbon neutral by 2030.[29] Alongside taking steps towards a more efficient fleet of aircraft that would use less fuel, Delta's carbon-reduction plan included the purchase of carbon credits generated from conserving rainforests, wetlands and grasslands. The legal challenge argued that the company's claim to be 'the world's first carbon-neutral airline' was 'false and misleading' because it relied on offsets that would do little to mitigate the impact of the emissions coming from the company's aircraft.

The court action came a few months after investigators claimed that rainforest offsets used by Disney, Shell and Gucci (among others) in their corporate environmental strategies were largely worthless as they were often based on stopping the destruction of rainforests that were not actually threatened in the first place.[30,31] While these attacks were dismissed by those running and using the offsets, they do nonetheless reveal the huge complexity of the issues and pitfalls linked with the effective implementation of such schemes, and the controversy includes very important dimensions linked with social equity.

When it comes to flying, it is worth noting that emissions are highly skewed towards the better-off. So extreme is the disparity between the rich and poor in terms of the pollution caused by flying that the wealthiest one per cent of the world's population account for half of all global aviation emissions. One study from Sweden's Linnaeus University found that in 2018 only 11 per cent of the world's population took a flight at all, with just four per cent of those journeys being overseas.[32]

I am not against electric vehicles or the principle of offsetting, but when it comes to the global deadlock linked with the disparities between richer and poorer members of global society, it is not only the current and historical patterns of consumption that fuel crucial divisions, aspects of the attempted solutions are also implicated. New technologies and environmental schemes can help, but they need to be done correctly; they should be based on social equity and linked with changed behaviour by the biggest polluters and consumers, rather than allowing them to just carry on as before, albeit with arguably greener credentials.

To this extent, steps towards cleaner driving and reducing what would otherwise be unavoidable emissions can obscure deeper challenges that must be addressed if we are to avoid ecological disaster. The simple fact is that to make human demands fit on our finite planet we are going to have to find ways to more fairly share the limited ecological space. Carrying on with high levels of material consumption among the wealthier minority and paying the poor not to develop in the manner the better-off have already done is probably not going to work as a long-term strategy – for reasons that are pretty compelling, when you think about it.

COMPENSATION

Discussions between historical polluters and those feeling the worst effects of environmental change are still hampered by disagreements. In 2007, the Bali Action Plan agreed by the countries of the United Nations referenced the idea of richer countries making loss and damage payments to compensate poor countries more vulnerable to climate-change impacts. This was 16 years after the concept was first tabled by the small island countries in 1991 prior to the 1992 Rio summit, proposing the establishment of an international insurance pool to help victims of rising sea levels, among other consequences of climate change that were placing such an economic strain on countries like the Maldives.

After the Bali meeting, it took six more years for a mechanism addressing loss and damage to be adopted. This was known as the Warsaw International Mechanism for Loss and Damage. The discussion went on from there, including at the historic Paris climate change summit in 2015, but even by then, 24 years after the original idea was first proposed, no money had changed hands. Discussions have continued, and at COP26 in Glasgow there were further talks about governance and an expert group on finance was set up to make progress, but still no money was made available. At COP27 at Sharm El Sheik in Egypt in 2022, the subject was on the agenda once again.

The progress made in Egypt was described as a 'breakthrough agreement' to provide loss-and-damage funding for vulnerable countries hit hard by climate disasters, with a few countries pledging finance. If it is to be effective, though, a lot of money will need to be transferred. According to the UNEP 2022 Adaptation Gap Report, the actions needed to cope with climate change, including everything from building sea walls to creating drought-resistant crops, could cost developing countries anywhere from US$160 billion to US$340 billion per year by 2030, swelling to as much as US$565 billion by 2050, if climate change continues to accelerate.[33]

A further breakthrough was touted at the Dubai COP28 summit in 2023, where more countries pledged money for loss and damage. A US$100 million allocation to the fund by the host country, the United Arab Emirates, was matched by Germany and slightly exceeded by Italy and France. The world's largest polluter, the USA, pledged just US$17 million and the second largest, China, only US$10 million.[34] In total, some US$700 million was promised, which is less than a half of one per cent of the bottom end of the range of damage costs estimated by UNEP to be needed by 2030. And while developing countries and campaigners insisted that this fund should be grant money, it is not yet clear if some of what has been pledged will be in the form of loans.

In 2024 at COP29 in Baku, Azerbaijan, a wider dispute about funding nearly caused the talks to collapse, as those suffering from poverty and the early impacts of climate change clashed with the richer countries over how much should be transferred to them from the nations that had largely caused the problem in the first place. It was yet one more reminder as to how the unequal starting points of nations is a huge barrier in dealing with pressing environmental questions.

So, after more than 50 years of global environmental discussions, it is undoubtedly the case that the huge inequalities that exist in both the source of the pressures and who is feeling the worst impacts of them have held back progress in avoiding (never mind reversing) further ecological decline. And it is, unfortunately, not only on the global stage where we can see that inequality has prevented us from making any real headway. A comparable dynamic has hindered movement inside many individual nations, including the UK.

8

False Choice

For decades, the disparities that have blocked progress on the world stage have obstructed action inside individual nations too. In the UK, there have been ups and downs. The greener future promised by the 25 Year Environment Plan introduced in 2018 under Theresa May's government was a moment of hope, setting out to fix many of our problems: reduce plastic waste, improve the health of our rivers, increase the abundance of wildlife and clean up the air, among other things.

At the time, I was working at WWF-UK, where we were campaigning to reverse the loss of nature. We welcomed the 25 Year Plan because it marked progress. We urged ministers to go further, however, including in their planned payment to farmers for providing environmental services, such as leaving uncultivated strips along the edges of rivers or improving habitats for birds. Many pioneering farmers were already doing some of this and the idea was to make it attractive for more of them to join in.

The upswing in ambition was driven in part by the emergence that year of a new protest movement that generated headlines and put pressure on the government. Extinction Rebellion pushed for far more decisive action on global heating, with dramatic and disruptive tactics that were not universally welcomed.[1] But in early 2019, when I took on a new role as the chair of Natural England, I thought such tactics were an important aspect of how overall momentum would be maintained. Bridges across the Thames were blocked by protesters, with colourful images broadcast on TV about the rising threats posed by climate change. These actions really pushed the issues up the agenda.

My new job was a major official appointment, involving working inside government to help move the ambitious focus on nature recovery embraced by the 25 Year Plan towards the actual delivery of positive change. Following my appointment by then Environment Secretary Michael Gove, I arrived for work on St George's Day 2019, having got off the tube at Westminster underground station, emerging by Big Ben and the Houses of Parliament, to walk to my new office across the road from the riverside Victoria Tower Gardens park that sits next to the House of Lords. I felt like an outsider coming in.

As I approached Parliament Square, I saw dozens of police officers surrounding the big plane trees. Extinction Rebellion protesters had climbed up them and were refusing to come down until the government agreed to do more to cut the pollution causing global heating. As I walked past in the spring sunshine, my thoughts were with those up in the trees. I reflected on my new role, advising ministers and Parliament, shaping decisions and the plans needed to get them implemented from the inside. Although in going to Natural England I had arrived at the other end of the spectrum from the arboreal campaigners among whose ranks I'd served for so many years, I was still in the same movement – only this time I would be seeking to make progress by using law and policy and presenting evidence-based advice to the government, rather than by means of ropes and banners.

A few weeks later, however, and the ropes and banners had made a difference. Theresa May decided to strengthen the Act of Parliament on climate change that I had helped to secure while I was director of Friends of the Earth, making it a legal requirement for the UK to achieve net-zero emissions by 2050. With that, the UK became the first major country to legislate for net zero, sending a signal well beyond its borders. It was criticized by the protesters (and many scientists) for not being enough, but despite the scientifically founded doubts, it was nonetheless a significant shift in gear and sent a powerful signal to businesses and investors as to the direction of forward travel. And the ambition on the climate agenda was soon to be joined by an uplift in the ambition to halt and reverse the decline of nature. In July 2019, Boris Johnson became Prime Minister, having won an internal Conservative Party battle for power, and among the many novel attributes he brought to that highest of political offices was a passion for environmental questions. This was in some part down to his father Stanley, who by then had been making the case for strong conservation laws for decades, including during his time working at the European Commission.

Boris Johnson's focus on environmental issues led his government to enact a range of new policies and laws, giving the 25 Year Plan legal teeth in a new Environment Act and adding impetus to the new sustainability focus on farming and fishing policies that the UK adopted after it left the EU. It was a good time to be leading Natural England, as doors opened, ambition expanded and budgets that had for years been cut to the bone were replenished; between 2019 and 2023, the money allocated to the organization increased threefold to over £330 million.[2]

When it came to the delivery of actual environmental improvements, however, many challenges remained. Questions of fairness and affordability began to bite in the wake of the COVID-19 pandemic, the invasion of Ukraine and a period of high inflation. This counter dynamic to environmental improvement was to unfold across a range of vital sectors, including food, water and energy. Let's have a look at each of those, and at how matters of fairness and affordability have influenced our collective ambition to fix environmental problems.

EATING THE PLANET

Before touching on the impact of our food system on our planet's life-support systems, I would like first to note how outcomes today contrast starkly with some of the doom-laden predictions made during times past. In the 1960s, some environmentalists predicted a kind of Malthusian crash in human numbers, as mass starvation came in the wake of rapid population growth outstripping food supply.

Not only did that not happen, but for most people food became relatively cheaper. Indeed, compared with 1957, and according to the Office for National Statistics, the proportion of average income spent on food more than halved, falling from approximately 33 per cent of wages, and then to 16 per cent 60 years later.[3] A similar pattern is discernible in other developed countries too, including the USA. In 1960, the average American household spent around 17.5 per cent of its income on food, whereas in 2019 that proportion had dropped to about 9.5 per cent. Serious volatility in food prices has since come in the wake of Russia's attack on Ukraine, with price rises for some foods made worse through the impacts of climate change, causing serious affordability issues for less well off consumers. Nonetheless, agriculture has managed to keep pace with rising demand, remaining resolutely focused on producing more in attempts to keep prices down. And as affordability issues bite, so environmental questions become less of a priority.

The scale of expansion in food production that took off during the mid-twentieth century is truly a massive achievement, although it has had serious environmental consequences. Indeed, it is our food system that poses the gravest threat to nature, while also being the second largest contributor of greenhouse gas emissions (after fossil fuels). The reasons for this hinge on both the scale of expansion in farming during the decades after the Second World War and how that has been accompanied by progressively more industrialized methods, based on more and bigger machinery, more land converted to farming, a stronger emphasis on monoculture, increased use of pest control chemicals, more fertilizer, more irrigation and more intensive and sophisticated plant and animal breeding methods.

The food revolution was also propelled forward through trade agreements that removed border taxes and other impediments to the movement of food between countries, while at the same time consumers became more used to and more willing to eat processed foods, rather than preparing meals from scratch, as their parents and grandparents had done. They also became, on average, relatively richer, with higher incomes than one or two generations before, thereby shrinking further the relative proportion of income spent on some essential items, including food.

Walking through well-stocked supermarket aisles loaded with products from all over the world dressed up in attractive packaging, it is sometimes hard to appreciate the scale of devastation that has accompanied our journey towards the cheap and secure food supply that is now enjoyed by many of the world's eight billion people. It is the case, though, that the factors which have driven the plummeting price of groceries are also the ones that have driven up environmental impacts, including declines in wildlife. Take the UK, where, during my lifetime, there has been a steady slide in the abundance of different species. It is a well-documented trend that is linked closely with more intensive crop and livestock production. Farmland birds are one group which has suffered terribly, with once common and widespread species, including turtle doves, corn buntings and tree sparrows, today hanging on with populations only a fraction of those seen in the 1970s.[4]

As well as telling a powerful story of their own decline, birds are also a useful indicator for other species, revealing a catastrophic scale of loss. One piece of research published in 2023 showed how, compared with the situation just a generation before, some half a billion fewer birds were present in Europe.[5] Researchers analysing a huge body of data collected by thousands of scientists across 28 countries for four decades found that it was above all intensive farming that was behind the steep population decline observed among the 170 bird species studied. While the overall decline in bird populations was equivalent to more than a quarter of the number present in 1980, the decline among farmland species was more than half, at nearly 57 per cent. By contrast, urban bird populations dropped by 28 per cent and woodland birds by 18 per cent. This study was particularly comprehensive, but it confirmed findings from repeated analyses going back years documenting how the rise of ever more intensive industrial farming is linked with the rapid and ongoing loss of wildlife.

In common with many consumers across the world, we have diets in Europe and the UK that are in large part reliant on a globalized food system that moves food over often vast distances, between fields, ranches and cages on one side and markets and consumers on the other. During the early 1990s, while searching for the rare Spix's macaw parrot, I travelled through central and north-east Brazil and the savannah woodlands known there as the Cerrado. Back then, the huge area of seasonal woodlands that lie to the south and east of the Amazon rainforests was under assault to make way for cattle pasture and soybeans, most of which were exported for animal feed. I saw fields that ran from one horizon to the other, with the red tropical soils that had until recently been covered with diverse wild woodlands converted to intensive crop production.

The beef and soybeans were produced to supply rapacious demand from international markets including in North America, the EU and China. This process of land conversion contributed to the loss of the woodlands, to wildlife decline and to carbon being released into the atmosphere. More than 30 years later, and that process of agricultural encroachment continues, eating ever further into the blocks of Cerrado that remain.[6] The soybeans travel across the world to intensive livestock units, from where they are converted by pigs, cattle and chickens into meat, and also vast quantities of excrement, which escape into the environment, causing serious damage, including to rivers, grasslands and other ecosystems.

Tropical forests in Asia, on the opposite side of the world, have also been under assault to supply different commodities, especially the burgeoning demand for palm oil. During the 1990s and since then, I travelled through vast areas that had been cleared of primeval forests and replaced with monocultures of palms that produce fruits rich in versatile oils used in a wide variety of products, including soup, bread, biscuits and non-food products such as toothpaste and cosmetics. It is a massive multi-billion-dollar business that has encroached upon millions of hectares of what was once the vast cloak of rainforest sprawling over New Guinea, Borneo, Sumatra and mainland Southeast Asia. Wildlife has been drastically diminished, with some species driven to extinction, while at the same time millions of tonnes of carbon have been released, especially where peat swamp forests have been cleared, drained and burned.[7]

Unsurprisingly, given what we know about the environmental impact of our food system, intensive discussions are under way about how farming can become more sustainable. Working as an adviser to food- and land-based companies, as a researcher supporting organizations promoting sustainable food, as a campaigner seeking to influence policies, and more lately as a public servant leading a government agency, I've seen close up many of the proposed solutions. They embrace a range of concepts summed up in words and phrases that include 'organic', 'regenerative', 'agroecological', 'nature-friendly' and 'sustainable farming'. While they vary in scope and priorities, all emphasize ideas that go in a broadly similar direction. These include using fewer or no pesticides, a focus on soil health (rather than large-scale fertilizer inputs), and increasing the extent of wildlife habitat, such as hedges, ponds and woods, blended into the farmed landscape, while diversifying production to be more mixed and less monocultural. For some globally traded commodities, various certification schemes have been established, including for beef, sugar and palm oil, to enable the sourcing of produce deemed sustainable. There have also been moves to link environmental health and the health of people through sustainable food.

All these shifts can have merit in reducing the impact of our food system on the environment, but because they sometimes entail short-term yield reductions, it's assumed that they will increase the price of food in the shops. Following the decades-long fall in the price of food, there has been great reluctance among producers, retailers, politicians and the public to put that trend into reverse, which means that farming methods with high environmental impacts continue to dominate global food production. Fears about the impact of higher prices on less-well-off consumers have been an especially important barrier to progress. The upshot of that is the perpetuation of environmentally damaging practices.

CHEAP FOOD

That cheap-food narrative is alive and well, and grows even stronger when stresses and strains cause prices to rise. Take the aftermath of the invasion of Ukraine by Russia in February 2022. Not only did this hit food exports from those two countries, in turn causing the prices of wheat and vegetable oils to spike, it also hit fertilizer markets, which relied on natural gas exports for the ammonium nitrate fertilizer at the very foundation of modern industrial farming.[8]

These factors soon became apparent in escalating shop prices, with the ensuing cost-of-living squeeze made worse by more than a decade of effective wage contraction, meaning that when food and other prices rocketed, families were hit harder. Indeed, productivity and wages had been stagnant for well over a decade, and in 2023 they were calculated to be back at the level they were at in 2005.[9] So serious was the impact of price hikes under these circumstances that even in a developed country like the UK it was not only the very poor who were thrust into crisis, it was also teachers, nurses and many others working in professional roles.

Plans then under way to move to more environmentally sustainable farming came under attack. There were calls in newspapers to focus on food production, not nature recovery, and to ensure that prices were kept down through increased food output and avoiding the cost of wildlife recovery.[10] Tim Lang, Professor Emeritus at London's City University and a world expert on food systems, has closely studied the interactions between food, society and environment. I've known Tim for decades and regard him as one of the most rigorous, cogent and informed authorities on food policy. I asked him about the extent to which policies linked with food were shaped by the plight of the least well-off, and what the implications of that would be for environmental policy.

'It is fundamental for the very simple fact that it's people on low incomes who are used as an excuse not to put the food system on an equitable and sustainable footing,' he told me. 'If environmental standards have to be raised, usually there's a cost implication of that, and always the counterweight that's thrown at attempts to improve standards is you can't do this because it will make food more unaffordable for the poor, when the poor are already badly served.' He then explained how the present food system was in large part shaped by a push to keep prices low: 'there was this broad agreement that making food cheaper was part of the rationale for intensification of agriculture'.

Lang reminded me that the UK was at the forefront of that agrifood intensification revolution, but that how, even in that country, major issues of food poverty remained. He pointed out how the clash between environment, costs and health justice that prevails in many countries is one reason why it has proved so difficult to arrive at policies and practices that not only produce good diets, but also look after the environment. He told me how when the chips are down it is 'price [that] has been probably the number one thing for politicians' and that is why they have been 'scared witless of being accused of deliberately wanting to raise food prices'. This is why, he added, 'there has always been some reluctance by the food policy power elite, big companies, some bits of the UN, leading G7 countries, to deliver integrated policies – always the threat that they might raise food prices weakens resolve'. He described what he called a resultant policy 'lock-in' that has limited environmental action. 'The evidence has been building up and up and up about the environmental and health damage from that model of agriculture and food production, [and it has left us] with low ambition,' he said, 'yet sorting out food would deliver multiple gains and common good.'

Our issues linked with diet are in large part down to the cheapest of cheap food coming to consumers via highly processed commodity crops, rich in sugar and fat, that in turn generate some rather expensive public health costs. In 2017, the prevalence of excess weight was 11 percentage points higher in the most deprived areas of the UK than in the least deprived.[11] In the most deprived tenth of areas, 67 per cent of people were overweight or obese, compared with 56 per cent in the least deprived. Children living in deprived areas were around twice as likely as children in the least deprived areas to be obese. Among children aged four to five, 12.4 per cent of those in the most deprived areas were obese, compared with 6.4 per cent in the least deprived. By age 10–11, this had risen to 26.7 per cent in the most deprived areas, compared with 13.3 per cent in the least deprived.

The link between poverty and poor diet is one of the factors that underlines some stark health differences across society. In February 2020, 'Health equity in England: the Marmot Review 10 years on' was published, following up on Professor Sir Michael Marmot's landmark report on health inequalities that came out in 2010.[12] The report examined the progress made in addressing health inequalities in England over that decade and found that the more deprived an area was, the shorter the life expectancy of the people who lived there was, while people in more deprived areas spent a larger proportion of their shorter lives in ill health, compared with those living in less deprived areas. The options open to poorer people in relation to food are part of that story, with reluctance to improve the environmental credentials of food production leading to an emphasis on cheap food, which in turn comprises the diet of the less well-off and makes a significant contribution to the poorer health outcomes experienced among those groups.

And what appears cheap in the shops comes with other huge costs for society beyond implications for public health. Take the huge amount of greenhouse gas emissions that accompany agriculture. The damage caused by extreme weather is sometimes also exacerbated by the effects of soil degradation, with compacted land depleted of organic matter causing water to run off, rather than be absorbed, making flooding worse. As we have seen, when it comes to climate-change impacts, and to poor diet, it is those who are most disadvantaged who tend to be the most vulnerable. It is thus even more important to question the logic of producing 'cheap' food and the extent to which it is really in the interests of the less well-off to promote such an approach to nutrition. The fact that people who rely on such food often have no alternative raises questions, or at least it should do, as to why that is the case, especially as this situation is evidently locking in more environmental damage.

Despite the widely held view that food production must be endlessly expanded to keep costs low, various analyses agree that we are already producing more than enough to feed everyone, and could do in the future too, as long as important caveats are embraced, linked, for example, with reducing food waste, providing a higher overall proportion of plant-based foods and limiting the amount of land used for non-food crops, including biofuels.[13] In short, the problem with food is not its availability, it is its affordability. And this depends on how much money people have to spend on it.

In some countries the proportion of income spent on food is at a historical low, and yet still a cost-of-living crisis has hit tens of millions of people – at the same time as a minority of the world's people have accumulated unprecedented wealth. What might the most sustainable solution be in this situation – relax environmental standards to push down prices a little further, or strive for more access to healthy diets, in part through greater equality? The first will be based on aspects that include reduced environmental regulation, the use of new pesticides and the clearance of more forests to make way for more fields, while the latter will require changes in social and economic systems. This is a real choice, with the first solution hastening the mass extinction of species and increasing global heating and the latter helping us to avoid this worst-case scenario.

A similar pattern is discernible in relation to water, where ambitions for environmental protection have been pushed down to maintain lower prices, ostensibly to protect less-well-off consumers.

LIQUID ASSETS

Fresh water is the life blood of the Earth's terrestrial biosphere. Without water, ecosystems collapse, and so do economies. Every facet of our lives depends on it, from farming to industry, and from our energy supply to the functioning of cities. It is unsurprising, therefore, that when countries began to put laws in place to protect the environment during the 1970s, it was protection of water that was one of the priorities for the new regulations and enforcement agencies. Progress has been made in some places, but despite many successful efforts to clean up rivers, including UK ones such as the Tees, the Tyne and the Thames which, during Britain's Industrial Revolution, were grossly polluted and effectively dead, the pressures on such ecosystems have nonetheless grown over time, and in some cases historical improvements have stalled or been reversed.[14,15]

Whereas once it was pollution from factories, from human waste, and from chemical works and food-processing facilities that killed England's rivers, today it is a different mix of pressures that drive major challenges. These still include historical factors, especially the impacts of sewage, but now on top of that there's the over-exploitation of groundwater, and the lower river flows linked with that. This is in part down to more homes being built in those areas where rainfall is comparatively low, such as around where I live in Cambridge, where some of the special chalk streams run dry as demand for water increasingly outstrips what nature can supply. With so much water pumped out of the ground, there is not enough left for the rivers.

On top of this increased demand, there are now the long dry spells and high temperatures caused by global heating, together with pollution coming from farming and the effects of centuries of land drainage that have desiccated landscapes, further impacting on rivers and wetland habitats. As is the case with farming, there has been a lot of talk about what needs to be done, and, similarly, some level of reluctance to do what is needed because of the perceived high costs of action.

In 1989, England's water supply and wastewater treatment infrastructure was transferred from public ownership to private companies. This was accompanied by a series of official regulatory processes governed by organizations that included Ofwat, which keeps watch on customer bills through agreeing how much the 17 water and wastewater companies in England can spend on the likes of maintaining sewage works, fixing leaks in pipes and building new water-supply reservoirs. Such work costs billions of pounds, and in the end it is paid for out of customers' water bills. Ofwat's job is to make sure that the bill-paying public get value for money through the water companies making proportionate and necessary investments. How this affects the environment is of course down to what is considered proportionate and necessary.

Should we invest the money needed to have vibrant rivers and wetlands rich in wildlife, or do we tolerate degraded and polluted waters with little natural value? The answer in many ways comes down to how much we are prepared to pay. Judgements about that are not by and large based on what wealthier people can pay, but are more strongly determined by the means of people with low incomes.

England's rivers are suffering from sewage pollution and low flow from our over-exploitation of groundwater because that investment in water infrastructure has been nowhere near enough to protect these vital aquatic systems.[16] Indeed, for the more than three decades that the privatized water companies have been in existence, it is the case that investment has been insufficient.

This is reflected in the distressing fact that only 14 per cent of England's rivers are deemed to be in 'good ecological health'.[17] This is the result of a combination of factors, and to be fair not all of them are linked with the water companies. A lot of the challenge arises from the slurry, pesticides, fertilizers and soils that run off the land from agriculture, together with the pollution coming off the roads. The water sector's contribution is, nonetheless, considerable, and includes not only sewage pollution, but also the huge amount of water taken from the environment to supply homes and businesses, which in turn deprives rivers and wetlands of vitality and health.

These pressures on water are one of the reasons why nature is in decline across the world. In England, such decline is evident in the fact that many of its internationally important wetland habitats, including the Norfolk Broads, the Somerset Levels and coastal areas like the Solent, have become degraded. In these and many other places the combined effects of pollution have caused drastic damage. In the summer, the coastal marshes and mudflats that lie along the Solent, a stretch of the south coast that is so important for shorebirds, are smothered with heavy, stinking mats of green algae which prevent the birds from feeding.[18]

This very visible environmental damage starts far up in the catchments of the rivers that empty into the English Channel, bringing pollutants that include the nitrogen and phosphorus that cause algae to grow more quickly. Some of this pollution comes out of sewage works (and, before that, household lavatories), and more comes from farming, as fertilizers and animal wastes escape from the land and into water courses. Some of the animal waste comes from farm animals being fed on soybeans (grown in South

America). All of this causes such huge problems partly because it is not being eliminated at source out of reluctance to push up customer bills for food and water.

Political emphasis on keeping bills low has placed pressure on regulators, including Ofwat, to slow down or avoid altogether the investments necessary to protect the environment. This view has been based largely on a cross-party consensus. For example, back in 2013 the Conservative government and Labour opposition both said that water companies should keep bills low to deal with what they saw as a cost-of-living crisis. Prime Minister David Cameron's spokesperson said he wanted 'to see household costs across the piece being reduced as low as possible. The intention is to try to reduce the burdens on hard-pressed families.'[19] In addition, he wanted 'regulators to look at the industry they regulate and make sure they are robust and delivering what they need to deliver for consumers'. Having worked as the chair of an environmental regulator, I know very well what such words mean – namely, that costs will not be increased.

In terms of the private water companies and their price regulator, Ofwat, this translated into a high bar for new infrastructure, such as more modern sewage treatment works or reservoirs that could hold winter rain and thereby take pressure away from rivers during the summer. These things cost a lot of money, and if there is a reluctance to pay for them, then pressures on the environment can mount. In 2015, in the wake of the political discussion that placed downward pressure on bills, the chairman of Ofwat at the time, Jonson Cox, signalled that he had heard the political direction very clearly, saying that households could expect to enjoy a 'decade of falling bills'.[20] Eight years later, in 2023, one of the main environmental stories in the media nearly every day was the sad state of England's sluggish and polluted rivers. I don't believe Cox's statement and that outcome are unconnected. His choice to focus on low bills and thus not to invest in environmental improvements led to the state of the environment deteriorating.

Government policies protecting people on low incomes from higher costs remain a major blockage. In the summer of 2023, when temperature records were smashed around the world and forest fires raged from Greece to Canada, England's water companies were advised by the government to slow down on investments that would help them to deal with climate change. Amid rising political concern about the financial impact of environmental targets, a letter was sent, under ministerial guidance, from the Environment Agency to the water companies asking them to ensure that their plans to deal with the impacts of climate change 'protect your customers from adverse bill impacts'.[21] With this guidance in mind, they were encouraged to gear their investments to prepare for the best-case climate-change scenario, which was cheaper than trying to anticipate the impacts that mid-range or worse-case scenarios could bring.

POWERING UP

During the 1990s, while working as the campaigns director at Friends of the Earth, I was engaged with efforts to create a cleaner and more efficient energy system. A major part of that was about using less energy, as well as making the case for wind and solar electricity to displace coal and gas stations. Making this shift was (and still is) resisted for a variety of reasons. One was because of concerns about energy security and the inability to generate enough power on windless days, or at night. Another was resistance by climate-change deniers, some of whom had links with the fossil fuel companies threatened by this transition. An additional prominent reason arose from concerns about the impact on consumer bills. This fear was harnessed by climate-change deniers as a tool to advance their in-principle opposition to action to stem the existential threat posed by global heating.

Thirty years ago, it was indeed the case that renewable power was more expensive than fossil sources, at least in terms of market prices. As with food, however, looking only at that aspect of the economic equation of energy and climate presents an utterly false picture of what is at stake, with the costs of inaction threatening to ramp up long-term costs far in excess of what is needed to shift our polluting energy system to clean alternatives in the shorter term. Damage caused by floods, storms, drought and sea-level rise threaten to cause costs far greater than that of decarbonizing electricity. This situation caused Sir Nicholas Stern, lead author of the British government's famous Stern Review into the economics of the climate crisis, to declare, 'Climate change is a result of the greatest market failure the world has seen.' Coming from the former chief economist at the World Bank, his words had a real impact.[22]

And yet, despite this and other warnings from economists (never mind those from environmental scientists), our ability to act has been seriously hampered, and not by a lack of insight or information, but by social conditions. The costs in the here and now drive policy choices, with the impact of short-term bill increases on people with low incomes pushing back against plans to cut carbon emissions, including measures to improve energy efficiency.

During the 1990s and early 2000s, Friends of the Earth worked with a coalition of partners in a campaign to do something about this apparent clash between environmental and poverty concerns through a drive for new laws to insulate the homes of those on the lowest incomes, thereby helping people to keep warm at a lower cost while also helping to meet environmental goals. The Campaign for Warm Homes sought to convince members of Parliament to back targets to end the phenomenon of fuel poverty, which afflicts people who can't afford to heat and light their homes. This was to be done through passing a new law that would mobilize investments to stop homes leaking heat. The initiative was led by Friends of the Earth's political campaigner Ron Bailey. Ron was adept at using parliamentary processes to get new laws passed and had been doing so for years, engaged in what amounted to political hand-to-hand combat.

As was the case with other initiatives he'd led, the idea was to get new laws to end fuel poverty into statute by using different parliamentary mechanisms, including private members' bills. These legislative vehicles start with a ballot process whereby each year backbench members of Parliament can put their names into a hat, and the 25 selected can introduce a new draft law to the Houses of Parliament (yes, it's bizarre, but true). Various constraints and rules make this quite a hard route through which to achieve the passage of legislation, especially without the backing of the government of the day, but it can be done, especially if you are in the top seven or so of the 25 MPs, as they tend to have more parliamentary time to complete the process of getting a bill turned into a law.

The campaign had been started in 1996 by Ron with other groups, including the energy poverty group National Energy Action, the Child Poverty Action Group, the National Housing Federation and the National Health Service Confederation. The latter was involved because of the clear links between energy poverty and health inequalities. Ron and his team worked out that the cost of a programme to insulate half a million homes per year would be about £18.75 billion, but that the positive impact on job creation, savings to the health service and reduced costs for homeowners would be about £22 billion.

The campaign was not, however, first and foremost about economic analysis, but about the social tragedy linked with fuel poverty. Stories were told of some of the individual victims of fuel poverty, including grandmother Mary Pike, who almost died of cold in bed in her own home. She'd developed hypothermia after her gas fire broke during a cold spell. When someone knocked on her front door, she managed to crawl to the door and was taken to hospital to recover. Another related to an old man who was effectively imprisoned in one room of his house. His home was so cold that he'd moved his bed into the single room he could afford to heat, with a blanket over the window and newspaper stuffed into crevices, and he sat next to a one-bar electric fire to remain warm enough to survive the winter cold.

These and many other cases inspired the campaign for new laws to make it a requirement for successive governments to do away with fuel poverty within 15 years. Letter-writing campaigns, constituency visits, parliamentary motions and public meetings were among the multitude of tactics used to galvanize support. A key breakthrough came several years into the campaign when, in

1999, Conservative MP David Amess agreed to sponsor a bill to do away with fuel poverty.

I worked with Ron to get more resources into the campaign and to refine the strategy as we went along, encouraging MPs to join key debates while urging others not to block the bill. It worked, and in February 2000 a new law to end fuel poverty was passed. A key clause set out the requirement to consult upon, and to adopt within one year, a strategy to end fuel poverty by November 2016. This was far from straightforward.

For one thing, the strategy that was adopted changed the definition of fuel poverty, eliminating at the stroke of a pen a million people from the category. It narrowed the definition so that it now included only those who expended at least ten per cent of their income on energy, rather than ten per cent of their *disposable* income. It also narrowed the definition to embrace only those households with vulnerable people, including the elderly, children and disabled people, removing a further million people from being classed as suffering from fuel poverty.

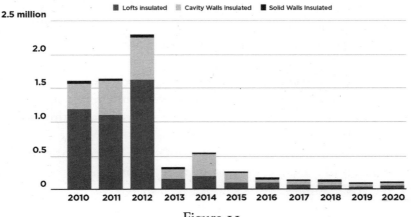

Figure 11

Ron explained to me how at first, despite this, the strategy went quite well, although it was set to suffer serious blows as money got tight. In 2009, the government went to court to argue that it didn't have a duty to end fuel poverty, as campaigners had intended the new law to mean, but only a duty to *try* to end fuel poverty. The judgement went in its favour, the political pressure on ministers was eased, and a more relaxed approach was adopted. Thirteen years later, when Russia invaded Ukraine, millions more people were plunged into fuel poverty as global fossil-energy markets were rocked by conflict and sanctions. Money that might have been spent in the early 2000s upgrading housing to more efficient standards would have saved not only bill payers but also the country many billions of pounds when the government had to intervene to support people who couldn't pay their fuel bills as energy prices went off any previously known scale.

A 2024 report prepared by Sir Michael Marmot's Institute of Health Equity for Friends of the Earth found that while there had been progress in improving the energy efficiency of the UK's housing stock, just over half of all households still lived in energy-inefficient homes.[23] This fact, coupled with elevated prices and poverty, meant, the report said, that more of the population was at risk of poor health and even death than had been the case a decade before. The number of people living in fuel poverty doubled between 2011 and 2023, with more than 12 million households spending more than ten per cent of their income, after housing costs, on energy, and nearly nine million in a similar position *before* paying their rent or mortgage.

Even in the aftermath of the considerable social impacts that followed the invasion of Ukraine, there are still voices that argue for retaining fossil energy as a core component in the energy system, and, remarkably, this position is deemed to be justified to protect the interests of poor people. We might wish to challenge that point of view not only in light of recent world events, but also given that in the years since the Campaign for Warm Homes, technological developments and economies of scale have led to the costs of solar and wind power falling below those of fossil energy.²⁴ This simple fact does still, however, sometimes have a low profile in the context of the very public claims being made by those sceptical about the desirability of moving to net-zero emissions.

In July 2023, the CEO of global oil and gas giant Shell, Wael Sawan, insisted that the world still 'desperately needs oil and gas'.²⁵ 'What would be dangerous and irresponsible is cutting oil and gas production so that the cost of living, as we saw last year, starts to shoot up again,' he said. He cited the circumstances of Pakistan and Bangladesh, which, he claimed, were unable to afford liquefied natural gas (LNG) shipments that were instead diverted to northern Europe because of the Ukraine war.

'They took away LNG from those countries and children had to work and study by candlelight,' he said. In making these remarks in this way he followed a familiar path of logic that invokes the interests of the poor to protect the utterly unsustainable status quo. Had we implemented in the UK ambitious home energy-efficiency programmes 25 years ago in the wake of the Campaign for Warm Homes, we wouldn't have needed so much gas for heating over the winter of 2022–23, and if Bangladesh and Pakistan avoid our mistakes and support poor people by making the shift to the renewable technologies that are both cheaper and less volatile in price, the dangers they face will diminish too.

Shell doesn't see it like that, of course, even though analysts pointed out how the idea of the choice having to be made between access to fossil fuels or working by candlelight was a gross misrepresentation of reality in a world where we know renewables are cleaner, cheaper and better for public health. In addition to ignoring the fact that renewables are often already the cheapest options, he also overlooked how Bangladesh and Pakistan are at the front line of climate-change impacts. The floods that hit Pakistan in 2022 cost the country tens of billions of dollars and directly led to the loss of more than two per cent of the country's GDP.²⁶ In Bangladesh, a one-to-two-degree increase in temperature will likely mean that up to 35 million people will be displaced by the resulting sea-level rise.²⁷

Shell, like other energy giants, is not in business to end poverty or to combat climate change. These companies are structured to maximize profits and will generally twist any narrative to continue to do that. If, however, their purpose as an energy company was instead about the provision of heat, light and mobility within environmental limits, then things might be different. For example, their focus would shift from petrol and diesel to building the charging infrastructure necessary to accommodate the rapid expansion of electric vehicles.

These same barriers of inequality that hold back progressive energy policies are delaying the changes that could make our transport systems healthier and more sustainable. One flash point arose in the constituency of Boris Johnson, the former prime minister who had brought such ambition for environmental issues to be at the centre of government.

DRIVING IN MY CAR

In 2023, a Labour policy intended to clean up London's polluted air by levying a daily charge on drivers of polluting vehicles led to their loss of the Uxbridge and South Ruislip by-election. The north-west London seat, held by Boris Johnson since 2015, was, contrary to expectations, retained by the Conservative Party. Many working people said they couldn't afford to pay the proposed charge, and this caused them to vote against Labour and for the Conservatives, who opposed the policy (even though it was originally their idea and first announced by Boris Johnson in 2014).[28]

The by-election opened a wider debate about the desirability of pollution-reduction policies, including the UK government's legally required goal to reduce greenhouse gas emissions to net zero by 2050. Prime Minister Rishi Sunak was moved to say that steps to avoid climatic catastrophe should be 'proportional and pragmatic'.[29] A chorus of calls followed from media commentators to reduce the pace and scale of ambition on environmental questions. Even the Labour Party felt political pressure to question how much British citizens should do to cut pollution.

The fact that senior political figures across the spectrum began to question not only measures for local air pollution but the whole net-zero strategy revealed the potency of the concept of fairness, or the perceived lack of it, as a reason to diminish environmental ambition. And this was despite estimates that in 2019 the equivalent of between 3,600 and 4,100 deaths in the capital were attributable to the effects of nitrogen dioxide and particulate pollution.[30] Also largely missing from the political narrative was the fact that improvements in air quality are associated with improved health outcomes, including reduced hospital admissions from childhood asthma, and that further cuts to air pollution would lead to significant reductions in coronary heart disease, stroke and lung cancer, among other medical conditions.[31]

Another dramatic example of such questioning came in the form of the *gilets jaunes* grassroots protest movement that began in France in November 2018. The popular outpouring of anger that came from this group was named after the fluorescent yellow hi-vis jackets that all motorists must carry by law in France, and which protesters wore as they occupied road junctions to express their opposition to sharp increases in the price of fuel. A hefty tax on petrol and diesel was a core element in President Emmanuel Macron's plan to make a transition to clean energy, but he was met with a turbulent reaction when it hit people on lower incomes.

The movement had sprung up from online petitions and was organized by working people, without an originating leader, union, established group or political party behind it. Following the first day of protests on 17 November, the actions continued daily with roadblocks, barricades and blockading of fuel depots among the tactics employed. Most of the hundreds of thousands who turned out to protest had jobs, with factory, IT, delivery and care workers among them. Their basic grievance came down to how their low incomes meant they couldn't make ends meet, leading to calls for the fuel taxes to be scrapped. It had a major political impact, causing the government to pause measures to cut climate-changing pollution. The feeling among many voters of being unfairly treated when policies arrive to cut pollution can, as we have seen, create political problems for elected representatives. It is not only in developed countries like the UK and France that a backlash can be triggered; the same thing has happened in other contexts too, including, during the last decade, in Nigeria, Ecuador and Chile.[32]

CULTURE WAR

The fact that inequality can lead to rejection of environmental policy creates opportunities to divide people for political ends. For example, when Donald Trump set out on his successful campaign for the US presidency in 2016, he garnered support in some large part from working-class voters for a platform that included his strong theme of climate-change denial. The conflation of climate scepticism with the concerns of key voter constituencies enabled Trump to win over people in West Virginia coal-mining communities, as well as Detroit car workers and the employees of Pennsylvania steel plants, thus enabling a billionaire climate-change-denying property developer to win the most powerful political office on Earth, with millions of votes from working people.

The deliberate provocation of a 'culture war' by pitching different groups against one another can garner political advantage. The opposition to climate policies among communities dependent on carbon-intensive industries is perhaps not surprising, nor that their fears of economic disruption, loss of identity and severed social bonds can be played upon for political advantage. Trump's victory was followed by the attempted withdrawal of the USA from the Paris Agreement on climate change, leading to global ramifications, as the world's largest emitter signalled its departure from the only plan the world had to avoid disastrous climate change impacts. The way the Uxbridge by-election result was seized upon as a springboard to weaken UK resolve had comparable qualities, with the sense of injustice deliberately stoked to create division. The more divided the society, the better this tactic works.

It was used again in September 2023 when, in the wake of a period of high costs and low wage growth, Prime Minister Rishi Sunak chose to make an attack on the UK's environmental ambitions. In a speech delivered from Downing Street he set out decisions to go more slowly on key low-carbon commitments – for example, in relation to how homes were heated, in the phasing out of fossil fuel powered cars, and even regarding the insulation of homes. He said that he was going to roll back on environmental ambition because he was concerned about 'unacceptable costs on hard-pressed British families'.[33] To strengthen the narrative, he pledged to scrap policies that were not even set to be introduced, such as sorting waste into seven different bins.

Former cabinet minister Jacob Rees-Mogg spoke up in support of Sunak's new approach, saying that he wanted policy 'that eases the burdens on working people', and that the new approach being set out was 'more pragmatic, proportionate and realistic'. This came in the wake of a summer of climatic disasters, including a catastrophic flood that killed thousands in Libya, but these two multi-millionaire politicians nonetheless cited the plight of people on low incomes as a core aspect of their case for weakening climate ambition. When he gave that speech, Sunak also chose not to attend the UN General Assembly taking place in New York, thereby sending a message to the global community that Britain was distancing itself from the international consensus on climate change that it had spent decades fostering. As the United Nations Secretary-General warned that humanity's impact on the climate had 'opened the gates to hell', shifts in UK policy oiled the hinges.[34]

Two days later, and nearly a quarter of a century after Conservative MP David Amess decided to back Friends of the Earth's warm homes campaign for better insulation, Rishi Sunak's government disbanded an energy-efficiency task force set up only six months before. Laura Sandys, a former Conservative MP and one of the members of the task force, expressed confusion, saying that energy efficiency must be the 'very first priority to reduce citizens' costs' and 'improve energy security'.[35] With such an obvious point being made, what could the motivation have been for such an apparently contradictory decision? In addition to sowing division for political ends, could it also have been an attempt to protect the interests of those who benefit from the status quo?

In January 2025 Donald Trump was back, returning to the office of President of the United States with even stronger backing that he had first time around. So divided and polarized was the USA that news of the 1.5 degree threshold of global warming being reached made no discernible impact on the debate, nor did the almost unbelievable scenes from Valencia in Spain, where massive damage to property and infrastructure, and the loss of more than 200 lives, came in the wake of a year's worth of rain falling in less than a day. Instead, climate change denial won the day.

FOLLOW THE MONEY?

The less well-off tend to live in more polluted neighbourhoods, suffer worse effects from environmental damage, be at the front line of climate-change impacts and have least access to green space, and yet these things are often perpetuated because the remedies to deal with them are, paradoxically, presented as unfair to the less well-off! At the same time, the business models that shape present circumstances and prove so immune to change, from our continuing reliance on fossil fuels to the illusion of cheap food and water, are highly effective at concentrating wealth. In relation to the industries I have briefly touched upon here – food, water and energy – it is instructive to note how approaches that diminish environmental progress, and that are justified in the name of less-well-off consumers, tend to be part of a way of doing business that accumulates huge fortunes.

Take the business models that lie behind industrialized farming: the processed foods that flow from it, and some of the fast-food and retail outlets that bring those foods to consumers, have proven highly lucrative. During the period of food-price inflation which followed the COVID-19 pandemic and the outbreak of war in Ukraine they have done especially well. Danny Sriskandarajah, the former CEO of Oxfam, told me about his organization's research into the rise of billionaires.[36] 'One stark finding,' he said, 'was the number of new billionaires added to the rich list on the basis of wealth gained largely through the food sector. We call them food billionaires, whose wealth has come in a large part from their ownership or shareholdings in food businesses. Twelve members of the Cargill family are US dollar billionaires and so this is just the sort of most vulgar manifestation of inequalities.'

The Cargill company's core business is in global commodity trade, buying from farmers and selling to food-processing companies that in turn manufacture the products we see on the shelves. It is one of the world's largest private companies and, according to Oxfam, was in 2017 one of four companies that controlled over 70 per cent of the global market for agricultural commodities. Those food-price shocks that have caused such problems for consumers were not such bad news for the likes of the Cargills, enabling that family to expand its collective wealth by 65 per cent since 2020, with four of those billionaires joining the Forbes list of the richest 500 people in the world.

Among other global giants that enable the system to function as it does are the likes of Nestlé, PepsiCo and Walmart, themselves extracting multibillion-dollar dividends for their shareholders. Dividends are the financial rewards paid to investors who've put their money into buying shares in profitable enterprises. The more shares they have and the bigger the profits of the company, the more money they make. And senior company executives have enjoyed handsome earnings too, resulting from a set of structures justified by, and protected in the name of, the 'cheap' food system that is the largest single engine of ecological destruction in the world today.

The drive for cheap water, too, has led to ecological damage on an epic scale across England. In the name of low bills there has for years been underinvestment in the infrastructure needed to protect and restore rivers and wetlands, yet vast fortunes have nonetheless been extracted from the water companies in dividends paid out to shareholders. England's water companies have been an attractive bet for investors from around the world – and why wouldn't they be, with captive customers supplied by regional monopolies providing them with what was until recently a low-risk home for their cash? Since privatization, England's water companies have paid out about £72 billion in dividends to their shareholders, while many of the executives running these businesses have taken multimillion-pound salaries and bonuses. In the decade prior to 2019, the privatized water companies paid £13.4 billion in dividends, while at the same time the directors who enabled this to happen, including by keeping investments to upgrade water treatment works low, saw their pay soar upwards.[37]

In 2019 alone, the income of the nine highest-paid directors rose by nearly nine per cent, with the two highest paid taking home more than two million pounds each. By contrast, the highest-paid executive at Scottish Water, which remains in public ownership, earned £366,000. And while vast fortunes have been extracted from monopoly utility companies providing an essential service, the companies themselves ran up huge debts – nearly £60 billion in total, with the largest one, Thames Water, accounting for about a quarter of that. It is worth remembering that when these companies were first sold off to private owners by Margaret Thatcher's government, they were debt-free, and on top of that, were set up with £1.5 billion of public money described as a 'green dowry' to help upgrade their assets.

The environmental challenges that many people believe the water companies should be addressing have not been adequately dealt with, and the money they borrowed was often not invested in new infrastructure but used to fund payments to shareholders, meaning that not only was pollution flowing down rivers, but cash was also flowing out of the UK to investors scattered from Kuwait to Australia. As is the case with our unsustainable food system, the political narrative behind this underinvestment in water infrastructure has been about keeping prices and bills low. But, at the same time, the water companies were motivated to increase shareholder benefits, with executives incentivized to do that via salary rises and bonus schemes. The pattern is repeated in relation to energy, where the perpetuation of unsustainable technologies has concentrated wealth, especially in the wake of the Ukraine war.

Despite the decades-long narrative that fossil fuels were necessary to combat poverty, the 2022 price hike not only made energy unaffordable for many consumers, but also led to an unprecedented increase in profits for oil and gas companies. This massive increase in income for these businesses was far from evenly distributed. One 2024 analysis estimated that in 2022 publicly listed oil and gas companies generated net profits of US$916 billion, with US companies between them making profits of US$301 billion. More than half (51 per cent) of this vast fortune went to people already in the wealthiest one per cent, through their ownership of the companies, including as shareholders. The bottom 50 per cent of US society between them shared one per cent of the money, and at a time when their wages were being destroyed by inflation, in turn largely caused by a sharp rise in energy prices.[38] It goes without saying that while the majority of people suffered in a cost-of-living crisis, the executives leading oil and gas companies were doing very well indeed.

For example, when, in 2023, the CEO of Shell came out to argue for more fossil fuels to combat poverty, he was quiet on another dimension of the overall picture – the fact that Shell had that same year filed the biggest annual profits in its 115-year history, surging to hit £32.2 billion in 2022, amounting to a doubling on the previous year.[39] His effort to secure rising dividends for shareholders was rewarded by an attractive salary package of £1.4 million, with performance-based annual bonuses of up to two and a half times that.[40]

In the case of the UK, there is a further dimension that lies behind these observations. It is revealed in an analysis estimating how, over more than two decades, the top one per cent of the country's richest people have been responsible for more emissions each year than the entire bottom ten per cent.[41] This context should perhaps lead to questions as to whether the policy choices made by societies are about protecting the common good, or whether they are more about protecting the tiny minority of the super enriched who do well out of food, water, power and transport being delivered in the way they presently are. I don't know the answer, but it is worth asking the question, considering the outcomes we're now seeing in terms of mass extinction and global heating.

The backdrop to our worsening climate and nature emergencies is this extraction of vast profits from economic sectors that are essential for human well-being, as well as the promotion of policies and business choices that perpetuate unsustainable impacts in the name of keeping key essentials, including energy, water, food and fuel, cheap. I am not saying these essential and basic needs should be more expensive, but I am arguing that we would be doing more to salvage our planet's declining life-support system if societies were less unequal. What if the vast billions accumulated by the tiny minority of super-rich people and organizations were instead more evenly spread and invested in solutions to the ecological crisis now upon us? Might that lay the ground for stronger ambition to be put behind sustainable farming, cleaner water, renewable power and better transport and housing? All these things need to change if we are to avoid the ecological disaster that presently we can't avoid (or that we are taking steps to avoid far too slowly) because of the vast economic disparities that exist in many countries.

One could go further in unpacking this situation, and observe how in some ways the false choice posed in the relentless programme to avoid environmental improvement in the name of lowering costs has itself been a key factor in perpetuating inequality. Without low prices, spectacular wealth would not have been so readily tolerated alongside poverty and need, and one way in which those low prices have been maintained has been through unloading costs onto the environment, with the resulting damage hitting the poorest hardest. The privatization of profit going hand in hand with the socialization of costs is a business model that thrives on inequality.

Rather than tackling inequality head-on, though, the emphasis nearly everywhere has instead been to foster more economic growth, with the assumption that the bigger the economy gets, the better off we will all be, including those currently least well-off. Let's turn to that next.

9

Growth

My many years of environmental work have taken me not only to remote rainforests, wild oceans, global summits and expert scientific meetings, but also to the UK's financial beating heart: the City of London. I've gone there to lead protests opposing financial institutions' support for ecological destruction, to speak at numerous conferences and events with insurers, banks and investors, and to advise some of the UK's most prominent companies.

Those organizations occupy imposing modern buildings squeezed between narrow streets where layers of history include the preserved remains of Roman temples (such as those in the basement of Bloomberg's European headquarters) and the mass graves of people who died centuries ago during the Black Death, some of which were recently unearthed during the construction of the new Elizabeth line rail link. The modern architectural wonders that dominate the UK's financial centre today include landmarks such as Heron Tower and the Lloyd's building and a range of eye-catching modern designs that have acquired colloquial names that include the Gherkin, the Walkie-Talkie, the Can of Ham, the Cheesegrater and the Helter Skelter.

Home to companies and institutions that are central to Britain's modern economy, they are physical manifestations of our moment in history. I sometimes ponder the extent to which such buildings sum up the era in which we live. I've been inclined to compare them with the Temple of Khnum in Egypt, with Europe's Gothic cathedrals, with Stonehenge, with the Great Mosque of Córdoba and with the Mayan temples, all of which speak of the worldviews of the societies that created them.

Those, together with many other surviving examples of sacred architecture, are reflections of religion and power and embodiments of the mindsets of the people who conceived and built them. Inspired by the patterns and proportions of nature, those ancient expressions of societal outlook speak of a time when people maintained deep philosophical and spiritual connections with the natural world. The imposing buildings that loom over the City of London today say something similarly profound about the idea that dominates our time. And while the buildings might be among that idea's latest physical manifestations, the history that lies behind how they got there runs very deep.

Our modern world came into being through industrialization while working in tandem with another strategy for economic expansion: colonialism. From the fifteenth century onwards, European countries embarked on the acquisition of overseas territories, with the mighty British Empire emerging as the largest the world has ever seen.

Its conquests dwarfed those of Rome, the occupied lands of Genghis Khan and the expeditions of Alexander the Great. At its peak it covered almost a third of the world's land, and while the sun has now well and truly set on Britain's imperial era, with its former possessions today run by their own populations, across the United Kingdom there remain glimpses of that past, and manifestations of what it evolved into. During my earliest days at school in the 1960s, the map on the classroom wall still had the territories of the British Empire shaded in pink. Since then, that colour on maps has mostly changed as more and more countries have gained independence, but some of the structures and processes that enabled the empire to thrive and concentrate wealth in the way that it did remain.

A walk along the docks in Bristol in the south-west of England reveals some of the layers of that global story. One chapter started with a small wooden sailing ship, a caravel, called the *Matthew*. A replica is moored on the docks today. Less than 25 metres long with two decks, one of them open to the elements, she had a crew of about 20 and could carry enough stores for seven or eight months at sea. You wouldn't think that the *Matthew* would be an adequate vessel to undertake intercontinental exploration, but that is exactly what she did.

On 2 May 1497, the Italian explorer Giovanni Caboto, who lived in England and was better known there as John Cabot, set sail from Bristol in the *Matthew*, heading west in search of a new route to Asia. Like Christopher Columbus, who, a few years before, set sail west from Spain to reach India, but instead found himself on the islands of the Caribbean, Cabot arrived in late June 1497 in what he called a 'new found land' – today known as Newfoundland, Canada. It was believed then to be the first journey from Europe to North America, but while Cabot was at first credited with the European discovery of that continent, he was subsequently found to be some centuries behind Viking explorers. In the eleventh century, pinpointed by recent archaeological analysis to the year 1021, they'd also reached Newfoundland, from settlements in Greenland. Irrespective of which Europeans had got there first, those arriving from England planned to come back, with Cabot's forays enabling Henry VII's Tudor England to lay claim not only to a new territory, but to an entire continent.

To European eyes, the new found land was a wilderness. With vast tracts of forests and untamed river valleys, it was virgin territory rich in resources, from timber to fish, minerals and animal skins, offering untold wealth to those who claimed it. So abundant was nature there that cod were taken on board the *Matthew* simply by lowering baskets over the side of the ship. Despite their sense of trailblazing discovery, Cabot and his men were not only 450 years behind the Norse travellers, they'd arrived roughly 13,000 years behind the Indigenous inhabitants of North America, whose ancestors had come from Asia across the Bering Strait on a land bridge that existed before sea levels rose at the end of the last Ice Age. Anticipating the possible discovery of hitherto unknown societies, Henry VII granted a charter for the voyage that set out permission to 'discover and investigate whatsoever islands, countries, regions or provinces of heathens and infidels, in whatsoever part of the world placed, which before this time were unknown to all Christians'.

In Cabot's new found land there were indeed people whose ancestors had lived there for a very long time. I found no sources to tell of any interactions that might have occurred between Newfoundland's Indigenous societies and Cabot and his men, but during the decades and centuries that followed there was to be a great deal of contact between Europeans and native peoples. For the original inhabitants of what we know today as Canada, and indeed the rest of the 'New World', it was to prove a far from positive experience. Eli Enns is a First Nations Canadian I met while attending global biodiversity negotiations in Canada in 2022. He is president of the Canadian section of the International Union for Conservation of Nature (IUCN), and is from the Tla-o-qui-aht First Nation, which is a Nuu-chah-nulth nation on the west coast of Vancouver Island.

He told me about how oral history tradition talks of first contact on the east coast: 'Everywhere I've met elders who all have a common story of early contact, of seeing the explorers as appearing harmless and in need of food with interesting things to trade, but also with deep concern that they would exploit all the natural resources around them.' Enns told me about a Cedar Man carving on what we would call a totem pole 'with one arm open in welcome and the other held back to remind each generation to be welcoming but cautious'.

It turned out that the Indigenous people of Newfoundland had good reason to be cautious. Descended from Algonquian hunter-gatherers, whose homelands extended across what are known today as Nova Scotia, Prince Edward Island and parts of New Brunswick, it was the Beothuk who inhabited the island of Newfoundland. It is generally considered that they became extinct in 1829, when Shanawdithit, the last known Beothuk, died in the town of St. John's, not far from where Cabot landed.[1]

Indigenous peoples were widely labelled 'savages' by Christian travellers, and early contact with Indigenous societies in the

New World led to a cascade of consequences, some of them deliberate (such as enslavement), and others accidental (such as fatal epidemics). In any event, the Indigenous societies were swept aside. Such was the confidence the Europeans had in their superior Christian values, and in the power and mandate that came from attachment to royal authority, that any consideration as to the fate of the native people was evidently nothing to occupy the attention of the new arrivals.

For centuries, the opening of new trade routes led to processions of ships sailing from around the world to Bristol's docks, offloading a vast array of goods to satisfy domestic markets craving sugar, spices, exotic textiles and novelty. From England's North American possessions, furs were the main export, causing the near extinction of the sea otter.[2] It was not only animals that suffered from the rapid expansion of trade. Much of the commerce was underpinned by and based on slave labour. Fortunes were created in Bristol, which saw the rise of spectacular new buildings erected during a golden age of Georgian architecture. The British Empire had by then become a truly global phenomenon, with, in 1788, a squadron of Royal Navy ships reaching Australia with the first colonists, thereby extending influence, connections and permanent presence to the far side of the planet.

Further along the dockside in Bristol is the SS *Great Britain*. The brainchild of engineer Isambard Kingdom Brunel, this revolutionary vessel was the world's first ocean-going iron-hulled steamship. A product of engineering and high-quality coal, the ship was very much a part of Britain's gathering industrial and imperial momentum. Launched in 1845, SS *Great Britain* was the first iron steamer to cross the Atlantic, making the journey between Bristol and New York in 14 days. She dwarfs Cabot's *Matthew*, towering above the tiny wooden ship which crossed that same ocean three and a half centuries before she did.

SS *Great Britain* provided the means to connect the world like never before, shipping people and goods from one continent to another, linking Britain's growing industrial might with the resources and new markets of empire. This generated yet more wealth for the owners of this new economy, whether it came from coal mines, new ships, plantations, timber and minerals plundered from newly occupied lands or the vast array of manufactured products flooding from coal-powered factories to feed rocketing demand for British-made consumer goods around the world.

The economy grew, and so did Bristol. Further along the dockside, a few of the cranes that once lined the city's quayside survive. These huge black structures that look as if they were inspired by the Martian machines in H. G. Wells's *War of the Worlds* brought ashore the shipments of grain, timber, minerals and manufactured goods, while loading into ships the outputs of British factories. Erected in 1951, they arrived just in time to mark the end of the empire, but not the end of the economy that it had helped to create.

Another symbol of that modern post-imperial economy is also moored on the dockside at Bristol, a vessel that embodies the contemporary version of global commerce. A sleek, shining, white multi-decked superyacht, complete with a small helicopter parked on top, *Miss Conduct* is an ocean-going ship valued at around £75 million, packed with technologies that Cabot's crew would have regarded as witchcraft. Capable of carrying more than 150 passengers and crew, the superyacht is a symbol of modern capitalism, and for oligarchs and billionaires a very visible and desirable exhibition of their extreme personal wealth. Those who have sailed on *Miss Conduct* include Donald Trump and George Soros.

Miss Conduct is a manifestation of how our economic system evolved. Beginning with expeditions to gain control of natural resources subsequently harnessed by the Industrial Revolution, it became spectacularly effective at concentrating wealth. The *Matthew* and *Miss Conduct* appear to come from different planets, but, moored next to each other in Bristol, they connect the present to the past. This is not least because the legal structures that have been so successful at generating fortunes to the point where an individual person can accumulate enough money to buy an ocean-going ship were first conceived in England. Initially designed to extract maximum financial benefit from imperial conquest, those structures today are at the heart of our modern growth economy.

EAST INDIA

While empire is often seen through the lens of war, conquest and cultural domination, for the British, the main motivation was trade and accumulating wealth. And while the initial claim of territorial ownership was under the flag of England, a different ensign soon fluttered above colonial possessions. With 13 alternating red and white stripes and the cross of St George in the top left corner, it was the banner of the first analogue of what we know today as a transnational corporation, and a forerunner of the golden arches of McDonald's and the Nike tick. It was the emblem of the English East India Company, founded by a royal charter granted by Queen Elizabeth I in 1600.

Like those modern brands, the English East India Company was a joint-stock organization, shares in which were bought and traded by shareholders who owned them without affecting the continued existence of the company, which was a legal entity in its own right. It became a very effective model for raising money for profit-making enterprise and rewarding those who risked their cash to make it happen through the payment of dividends derived from the profits. A royal charter reduced the risks to shareholders, who, in the event of the company failing, would have limited liability and thus would avoid ending up in a debtors' prison.

Other state-backed joint-stock companies followed, including the Dutch East India Company in 1602. The Hudson's Bay Company was founded by royal charter in 1670 for the exploitation of the natural riches of Britain's vast imperial territories in North America that were first opened up via the voyage of Cabot in his tiny wooden *Matthew*. The new business models hastened the exploitation of the natural environment. Before the invention of modern fabrics, beaver fur was hugely popular for making waterproof hats, which the Hudson's Bay Company supplied. Beavers had been wiped out across Britain during the 1600s, and across much of Europe too. Market demand for beaver skins remained very strong, however, and the Hudson's Bay Company grew hugely successful, causing the population of these animals to crash from about 400 million in the 1500s to near extinction by the 1850s.[3]

Another attempt to harness imperial possessions for trade was seen in the establishment of the Imperial British East Africa Company. With the British government reluctant to provide the resources for the administration of East Africa, this new commercial entity was established to develop trade in the areas in that region that were controlled by Great Britain, covering a vast tract stretching from the Indian Ocean coastline of what is modern-day Kenya to the far side of Lake Victoria in what is now Uganda. The company was incorporated in London in April 1888 and granted a royal charter by Queen Victoria a few months later, at which point it assumed responsibility for that vast expanse of Africa. An early example of privatization, the company took on the not inconsiderable task of asserting its authority over the native peoples who were in its sphere of influence.

Successful capitalists today, in common with those operating in centuries gone by, make profits by investing their money in ventures they judge to have a good chance of generating a return. For example, investors in the Hudson's Bay Company put money into the purchase of ships and other equipment needed to catch, process and transport beaver skins for urban consumers, generating a handsome profit in the process. A wrecked ship might lead to expensive losses, but insurance was another invention of early capitalism that enabled entrepreneurs to risk pool, thereby avoiding the potentially ruinous effects of bad fortune.

The existence of insurance encouraged investors to back businesses that might otherwise be deemed too risky, although the bigger the gamble, the higher the cost of the insurance premium. The legal status of limited liability companies also helped to manage risk, and these became a common vehicle for mobilizing investment and for injecting the vast amount of capital needed to finance not only empire but also the Industrial Revolution. London coffee houses became hubs for merchants and share dealing, founding the financial district that is today the City, with its modern buildings and its annual throughput of trillions of pounds' worth of trade.

A successful investment in one venture could generate returns that could then be ploughed into another, accumulating further wealth for the capital owners. This required skill, knowledge and networks, and it was aided by political policies that encouraged investments to flow by increasing the rewards for those allocating the capital, or those who ran businesses – for example, by granting them rights to exploit particular territories.

These limited liability companies proved to be a highly lucrative means for generating wealth, so they grew, backed by the governments that were closely associated with the well-connected and wealthy elites who ran them. The expansion of wealth meant that tax revenues increased, so governments had good reason to back such enterprises. Until the nineteenth century, the privilege to set up such companies was granted by the Crown, but so successful was this means of harnessing capital that the idea spread, and today the model of limited liability companies thrives and is a vital pillar of the global economic system.

In the white heat of the Industrial Revolution, major landowners began to harness their wealth to build new mills and factories, initially located on rivers for access to water power, but branching out across the country more widely with the advent of steam. It was not only financial capital, natural resources and technology that were involved, but also much of the labour force that once worked in agriculture but moved to towns and cities to be employed in factories. And alongside those who were paid there toiled an army of slaves who, through violence and coercion, had become the backbone of the sugar economy and vital to producing the cotton so central to Britain's industrialization. Vast wealth was generated, mainly for the factory and plantation owners.

It is difficult to chart the size of economies through history, not least due to the sketchy nature of the available data. There is, however, broad agreement as to the slow pace of change in the scale of economic activity over time. For centuries, the global economy barely grew at all, varying mostly due to changes in the population. The more people there were, the bigger it got.

Before widespread industrialization, economic systems used to work largely within fixed boundaries, defined by the productivity of the land and the number of people it could feed using human and animal labour. This changed dramatically with the onset of the Industrial Age. In Great Britain, per capita economic activity expanded by about 40 per cent between 1700 and 1800, and more than doubled between 1800 and 1900. As the process of industrialization and global economic integration (including via colonialism) took off, so did the pace of expansion. According to the International Monetary Fund (IMF), between 1900 and 2000 the world economy increased about nineteenfold, driven by an average annual rate of growth of three per cent.[4]

So it was that the arrival of new technologies and the conquest of faraway lands transformed economic possibilities and lifted the limitations to wealth creation set by the constraints of land and labour. A combination of coal and empire smashed historical boundaries, enabling unprecedented economic expansion. And as this expansion gathered pace, so the question arose of how to measure it.

ONE NUMBER

I'm not an economist, but considering how economists have over the years repeatedly shared views about environmental questions, it only seems fair that environmentalists should return the compliment. Let me start by saying that as a lifelong environmental advocate I have reached the conclusion that shifting the fundamentals of our modern economic system is vital if we wish to address environmental conundrums. One of these fundamentals concerns the notion of gross domestic product, or GDP, and, more importantly, the policies adopted to grow it.

GDP is the most widely used measure to chart economic expansion. It encompasses the total cost of goods and services sold in a year, and was put forward during the 1930s as a means of tracking the size of an economy and how it changes over time. Despite its originators warning that it should not be used as a proxy for overall progress, that is exactly what it has become. Rather than being simply one measure of how much stuff is being consumed, GDP has become a focal point for policy and a bellwether of national advancement, with our economic system now dependent on GDP growth. For example, the financial system relies on it to calculate the returns needed to repay debts and to meet the demands of investors who desire a return on their capital. Governments need growth to generate the bigger tax revenues needed to pay for health, education, police and all the essential machinery of society.

GDP has become an obsession for many influential organizations, from political parties to major companies, and from global economic institutions to central banks. For them, achieving more GDP is the overriding priority. All else is subservient to it, not least because it is assumed that if GDP is expanded, we will automatically achieve other desirable outcomes. Indeed, the political narrative that accompanies the drive for GDP growth around the world is generally attached to the idea of increased well-being for people being attained via increased access to essentials such as housing, healthcare, sanitation, education, power and food. While this is true up to a certain point, the fact that we live in a world of unprecedented environmental emergency and, for modern times, unparalleled inequality should cause us to pause before declaring higher GDP as an unambiguously positive goal.

But how did we get to the position we're in now, where the single-minded pursuit of GDP growth goes almost unchallenged in mainstream debate, despite the scale and urgency of the social and environmental challenges we face? Over the years, I have pieced together the history of ideas that led to this present moment, and sharing a little of that here will help shed some light on how we got to where we are, and, I hope, lay the ground for logical ways forward.

GDP was first proposed as a single unified measure of economic change by an economist called Simon Kuznets. He was born at the turn of the twentieth century in Pinsk, a city in the Russian Empire (now in Belarus), and studied economics and statistics at the

Kharkiv Institute of Commerce (now the Simon Kuznets Kharkiv National University of Economics in Ukraine).[5] When Lenin's Red Army prevailed following years of civil war and began to lay the foundations for the Soviet Union, Kuznets joined thousands of fellow citizens in emigrating to the USA. He gained a PhD in economics at Columbia University and secured a position with the National Bureau of Economic Research, a respected economic think tank.

His work sought to understand the Great Depression and how the economy went from a healthy and apparently stable position to one in which fortunes were lost and mass unemployment suffered. The pre-depression context was one of extremes, in which a handful of individuals, such as oil man John D. Rockefeller, banking magnate John Pierpont Morgan and the steel giant Andrew Carnegie, controlled colossal wealth while most citizens experienced extreme precarity, dependent on the wages they could earn day by day to survive. Money poured into an ever-rising stock market, even though there was no comparable trend in the real economy of actual businesses producing products and services, highlighting a disconnect between financial speculation and what was really going on. In October 1929, the collapse came and set in motion events that would sweep across the world. Credit markets dried up, unemployment rocketed, families stopped spending and protectionism mounted as a crisis deepened that would last until after the Second World War.

In seeking to better understand what had happened, Kuznets wrote a report called *National Income, 1929–1932*, which captured in a single measure all the economic production across the entire country, from individual citizens, companies and government.[6] It would be expected to go up in good times and down when things were not going so well. GDP was born. The tool that Kuznets put forward was intended to enable policymakers to better understand broad economic trends. It was a breakthrough, allowing the economic strength and complexity of a huge country to be summarized in one number. This simplicity helped to make it a priority. Kuznets warned that despite the obvious attractions of relying on a single measure, it was not necessarily a good tool for wider decision-making. History reveals the extent to which his warnings went unheeded.

Even as the Second World War raged on, a pivotal meeting held at Bretton Woods in New Hampshire in 1944 began to set out the architecture needed for post-war economic recovery. Institutions that remain with us today, including the World Bank and the IMF, were established. As part of the package, it was agreed that during the period of reconstruction, GDP would be the main tool for measuring countries' economic performance. With that, the measure went global and became an international metric against which countries would not only measure themselves, but also be compared with each other.

During the decades that followed, scepticism was sometimes expressed as to the extent to which GDP growth could accurately measure the well-being of societies. In 1959 American economist Moses Abramovitz cautioned that 'we must be highly skeptical of the view that long-term changes in the rate of growth of welfare can be gauged even roughly from changes in the rate of growth of output'.[7] He was a rare sceptic, but not alone. In March 1968, US Senator Robert F. Kennedy delivered a speech that laid out serious doubts as to the wisdom of pursuing GDP growth as an accurate proxy for social well-being.[8]

'We will find neither national purpose nor personal satisfaction in a mere continuation of economic progress, in an endless amassing of worldly goods. We cannot measure national spirit by the Dow Jones Average, nor national achievement by the Gross National Product. For the Gross National Product includes air pollution, and ambulances to clear our highways from carnage.' He went on to point out how it 'counts special locks for our doors and jails for the people who break them. The Gross National Product includes the destruction of the redwoods and the death of Lake Superior. It grows with the production of napalm and missiles and nuclear warheads.' Despite these powerful interventions, full vindication for Kuznets's ideas came in 1971, when he was awarded the Nobel Memorial Prize in Economic Sciences.

A few prominent figures have continued to raise questions, including Sir David Attenborough. Speaking at an event at the Royal Geographical Society in London in 2013, the veteran naturalist and broadcaster pointed out, 'We have a finite environment – the planet. Anyone who thinks that you can have infinite growth in a finite environment is either a madman or an economist.'[9] But despite sporadically expressed misgivings, GDP's high profile has remained constant, with the assumption that countries should keep expanding their economies, and that if they do, everything else, including full employment, high standards of healthcare and universal access to education, will naturally follow.

GDP has become the statistic to end all statistics, a single number to trump all others. It was adopted as the headline commitment of governments across the world and turned into a league table, with those at the bottom lambasted by commentators, ridiculed by political opposition and scolded by expert institutions. Thus, out of the social and economic carnage of the Great Depression and the global conflagration of a world war rose the ultimate measure of a country's overall welfare – a window into an economy's soul. Those questioning the sense of using GDP as a single central measure of the policies adopted to make it grow were increasingly cast to the fringes of debate. When was the last time you heard a political leader, top economist or chief executive make the kinds of points shared by Robert F. Kennedy more than 50 years ago, or by Sir David Attenborough more recently?

Yet, despite a massive body of evidence showing that a singular focus on GDP obscures other vital trends and leads to an underestimation of socioeconomic inequalities and environmental degradation, it has reigned on. Across the board, from the right to the left of politics, in nearly all the major countries, there is a near seamless consensus. Editorials in the main newspapers rarely dabble with the occult activity of questioning the wisdom of having GDP growth as our headline indicator, and (with rare exceptions) neither do the leaders of political parties. The only questions that are really debated concern *how* to achieve growth, and on that subject, there are basically two broad schools of thought: they can be summed up as Keynesianism and neoliberalism.

TWO IDEAS

During the post-war years, the ideas of British economist John Maynard Keynes were to become prominent. His major intervention came in 1936 with the publication of *The General Theory of Employment, Interest and Money*.[10] It was, like Kuznets's work, an attempt to understand the Great Depression, forming the theoretical basis for the kinds of interventionist policies that Keynes favoured, whereby, he argued, government expenditure on infrastructure, housing and key industries would put fuel into the economic tank in ways that would drive demand, and thus growth.

He was highly sceptical of the British government's austerity policies adopted during the economic slump of the 1930s. Even if governments had to borrow large sums of money, his strategy was still sound, he said, and would prevent business losses from being so severe during an economic slump as to bring the economy to its knees. His idea was that a fiscal stimulus in the form of an expanded government spending plan would lead to more business activity, and more investment and spending, thereby boosting output, generating more income, and leading to GDP growth that could be even bigger than the initial stimulus.

Keynes's ideas became influential and were important in shaping post-war economic policy, when there was strong recognition of the role government could play in increasing demand through borrowing and spending. Keynes's proposals contrasted with what he called classical economics, a broad school of thought dominant during the eighteenth and nineteenth centuries which is often traced back to the ideas of Adam Smith. Smith was an eighteenth-century Scottish philosopher whose seminal work *The Wealth of Nations* was published in 1776, just as the Industrial Revolution was starting to get under way in Britain.[11]

Smith's basic point was that wealth is created through productive labour motivated by self-interest guiding people to make choices about how to put their resources to best use, thereby leading capital investments to naturally flow to where the most profit could be returned. He was the originator of the idea of the 'invisible hand', predicated on the notion that allowing individuals to decide how to use their land, skills, money and facilities as they saw fit would generate the most benefit for the nation, and that this pursuit of self-interest would result in a self-organizing system that would be better for everyone.

Both of these broad schools of economic thought have been tried. For a time, after the Great Depression and the war, there was a period of government-led economic planning. It was then that Prime Minister Harold Macmillan proclaimed that the British people had 'never had it so good'. During the 1970s, however, an influential counterview gained momentum. The US economist Milton Friedman was the most influential advocate of free-market capitalism during the twentieth century, making the case for a full-on return to classical economics, an idea which, while controversial and radical at first, began to gather momentum. In the 1970s, political leaders were making the case for adopting his mix of policies to unleash free-market forces to drive economic growth.

Core to his theory was a theme now known as monetarism, whereby, in contrast to the Keynesian emphasis on spending, the supply of money was tightly controlled to prevent runaway inflation. With the role of government dramatically shrunk, the economy would be free to operate via Adam Smith's 'invisible hand', with capital free to move where investors and businesses saw most prospect for profit, thereby driving growth. Friedman's ideas favoured small government, deregulation of most areas of the economy, lower taxation, policies to promote free trade and a shift of state-run entities into private ownership.

One fundamental consequence of Friedman's vision would be a tendency for power to move away from democratically elected political institutions and towards market actors. In economies with diminished regulation, they would be able to make more decisions without recourse to the common good. This small-government approach, driven by market forces and a rejection of central planning, had some foundations in classical economics, but between the 1960s and 1980s it evolved further into a political philosophy. We know it today as neoliberalism.

The moment when this political ideology took hold is not easy to pin down, but some specific events stand out. The election of Margaret Thatcher as UK prime minister in 1979 and that of Ronald Reagan as US president in 1980 are generally cited as key turning points. The thinking that had been explored at the 1972 Stockholm environment summit was invisible in their campaigns for office, but both emphasized free-market policies, the shrinking of the state and a bigger role for private companies. For Thatcher, it was trade unions that needed taming, and she thought that old, inefficient industries had to be left to die or be killed. For Reagan, it was a balanced budget that was key, together with a loud call for ambitious plans for growth at a time of stagflation (low growth and inflation).

It would be wrong, however, to believe that the rise of Thatcher and Reagan led on its own to the prominence of neoliberal ideas. The truth is that it was the result of a long, well-organized and orchestrated campaign backed by various entities, including a group called the Mont Pelerin Society. Its first meeting took place in 1947 at the Swiss village of Mont Pèlerin, close to the town of Vevey. That initial gathering was convened by political scientist and economist Friedrich Hayek. He invited 39 scholars, mostly economists, together with leading historians and philosophers, to meet at the Hotel du Parc to discuss the future of liberalism.

Mont Pelerin Society member Milton Friedman sat at the intellectual knee of Hayek, sharing his views about the need to resist government intervention in a rapidly changing world where the political trend was then towards socialism. The society continues to meet to this day, advocating free-market economics and the transfer of functions provided by government to the private sector. During its history, the society has advised Ronald Reagan, Margaret Thatcher and other leading neoliberal political figures. Its members have also set up other neoliberal think tanks, such as the Institute of Economic Affairs in the UK and the Manhattan Institute for Policy Research in the USA.

Think tanks and advocacy groups like these were an essential component in the mix that led to the rise of neoliberal ideas, but events also played a vital part. During the presidential election campaign that led to the election of Ronald Reagan it was far away in the deserts of Iran that the political demise of Reagan's election rival, President Jimmy Carter, would be sealed.

Carter was by any standards a green president. He was briefed about climate change and believed it was a real threat, even back then. Carter set out a decidedly Keynesian policy to build a programme to promote solar power in the USA. In 1979 he installed solar water heaters on the roof of the White House. He was ahead of his time on environmental questions, but was set to lose office in large part because of how a hostage crisis was to pan out. The USA had been humiliated by a hostile Iranian regime and by a group of anti-American students who'd seized the US embassy in Tehran in November 1979, taking 53 diplomats and citizens hostage. Carter decided to stage a rescue in April 1980. Unfortunately, a series of disasters led to the deaths of one Iranian civilian and eight American service personnel, leading to even more American humiliation. Carter went on to lose the presidency to Reagan in November of that year. Some observers see this moment as pivotal in explaining how government-led attempts to reduce climate-changing emissions were abandoned in favour of small government, deregulation and pro-free-market-growth policies.

By 1992 and the Rio summit, the narrative of economic growth via free markets, private enterprise, small government and deregulation was fully established. Its advocates presented it as a rising tide that would lift all boats, and aligned their pro-growth programmes as the only route to cutting poverty and achieving sustainable development. The neoliberal economic revolution was seen by many to have prevailed, not least in the titanic struggle between East and West, with the collapse of the Communist Soviet Union and the Warsaw Pact serving as the ultimate vindication for the western growth model. As economic historian Simon Szreter explained to me, 'The breakup of Eastern Europe and the Soviet system produced this sort of euphoria that liberal markets are here to stay, and that everything's wonderful and we can turn this model into the entire global economic idea. All we need is deregulation.'

This in turn bolstered political motivation for a new world trade organization, to facilitate the removal of barriers to the movement of goods and services across borders. Listening to negotiations at the Rio summit, I was struck by what a major talking point this was. Leaders of western nations advocated that sustainable development would come from economic growth, which in turn would be derived from more global trade. With the threat of Communism removed, there was a new confidence in creating a free hand for wealth and a far more relaxed attitude towards the rise of extreme financial elites. This economic globalization of course came with many consequences, including in some countries the closure of heavy industry, such as at Teesside in the north-east of England, as companies searched for cheaper places to carry out production.

Veteran environmental journalist Geoffrey Lean suggested how things might have gone had a different economic narrative prevailed during that critical period. He reflected on the missed opportunities that came with the demise of Keynesian economics. 'When one economic philosophy dies people start thinking up a new one and they reached around for what there is, and what was there was Milton Friedman and we went straight down that route.' He recalled how others were working on a kind of 'green Keynesianism' and how they sought to see that embraced at the Rio summit. It didn't come to anything, but if it had done the world would now be a very different place.

Irrespective of the economic ideas that GDP measures, ever since its widespread adoption there have been several revisions to its basic construction. One of the most significant came in 1993, when finance was explicitly included as a productive part of the economy. As the economy has grown, the proportion of it accounted for by finance and banking has increased as well. Before these sectors were added, they were considered a neutral part of the growth project, the oil that lubricated the engine, but now they are parts of the engine in their own right. Wall Street and the City of London became central to growth, with the vast fortunes parcelled out in salaries and bonuses driving further economic inequality.

In the USA in particular, the neoliberal growth machine became equated with freedom, and thus was seen as even more of a political counterpoint to the authoritarian regimes of the collapsed Soviet bloc. It has also been linked with God and Christianity, creating an alignment between faith and the economic system. Indicative of the connection is the slogan that has since 1957 been printed on the back of the US one-, five-, ten-, twenty-, fifty- and one-hundred-dollar bills: 'In God We Trust'.

This history lies behind our modern context and the pathways that have caused us to arrive at the system we now have. That system has evolved through choices that began with industrialization and colonialism, and has aligned over time with notions of growth embedded into the very DNA of our political debate. Indeed, so embedded has the system become that it is not for the most part debated at all. Yet as we face the huge challenges posed by rapid environmental degradation and rising social inequality, debate it we must.

One person who has been doing just that for some time is Professor Tim Jackson. He leads the Centre for the Understanding of Sustainable Prosperity (CUSP) at the University of Surrey. He comes to the daunting subject of how to match human demands with the capacities of our finite Earth as a former member of the UK government's Sustainable Development Commission. He is also the author of the widely acclaimed book *Prosperity Without Growth*.

Jackson sees good reasons why GDP growth has become so deeply intertwined with our economic thinking. One relates to the basic reality of rising living standards. 'It measures what people buy, and the more people are having what they want or need, that's progress. Society is better off because people have the option to pay for the things they need and want.' On top of this basic rationale, there are also structural reasons for GDP growth having become so bound up in the economy. 'Once you build it into your system, it's very difficult to get off as we become growth dependent. You can't make the financial markets work and can't make your public welfare budgets work.'

Jackson also points to the changing demographics in many western countries, whereby increasing longevity means there's a declining proportion of working-age people. 'We've got more life years and a relatively smaller working population to support an older population. The way to make that work is to get some growth.' He says this adds up to 'a mix of reasons why growth becomes so fundamentally important in the economy'.

Another reason why GDP has become so deeply embedded, he says, is because it is actually a very useful measure for understanding the economy. 'Because it has three triangulated components, one around the value of economic activity, one around spending and the other income, with those three different measurements, GDP is actually a very good way of figuring out what your economy is doing. It tells you a lot about your employment ratios, how many people you need, what your public service expenditure and your dependency on taxation is. There are all sorts of things you can do with that measure, and that's one of the reasons why it remains so powerful, even though there are lots of assumptions within it that are not entirely objective scientific assumptions.'

While seeing the rationale for GDP growth, Jackson also recognizes some major drawbacks. One of them relates to the environmental impact. 'The more GDP you have, the more economic activity there is and the bigger the impact on the planet. This means that there's likely to be less GDP in the future because at some point you start stepping beyond the planetary boundaries, and that feeds back to hit agriculture and so on.' He adds how this omission of the environmental impact of growth 'is a very good reason for saying that this indicator is not good enough any more, because it has no measure of that impact'.

Alongside this serious downside, he has doubts about the extent to which headline figures about GDP work as reliable proxies for social well-being. 'It's a measure of the economy as a whole, and that doesn't tell you anything about the distribution of the income in society, or the potential that people have to achieve decent lives.

And we know that because as overall income has been growing, so has inequality. For the poorest people in society, typically their incomes have over several decades stagnated or fallen.'

He adds that this is massively important in terms of the argument that GDP growth automatically supports social welfare. 'You cannot simply translate an aggregate or an average GDP into the welfare of the population, particularly if you've got a very large number of people whose incomes, welfare and well-being are all going downwards, who have much less access to goods and services and who suffer from poor health as a result, and who have worse environmental conditions and worse housing […] The average doesn't equal the aggregate.'

So GDP tells us about overall consumption, but not about well-being. It speaks of production, but not of the pollution that comes with it, or of the depletion of essential natural resources as consumption goes up. It also fails to measure the inequalities that can be exacerbated by the choices made to grow GDP, which we have seen are so fundamentally linked to the climate change and ecological declines now so visible across the world. Despite all of that, though, it nonetheless remains the most powerful idea shaping outcomes everywhere. Such has been the success of its advocates, using simple ideas and connecting notions of growth with freedom, choice with liberty, and even God with progress, in embedding an idea that is now barely questioned, certainly not by those who wield most political, economic or cultural power.

DRIVERS OF INEQUALITY

The rise of the company structures that generate growth and returns for investors is one of the factors that explains the income differences between those at the top and bottom of many societies. Income derived from wage labour is qualitatively different to that coming from capital and from the ownership of income-generating assets, such as company shares, land or property. People in employment can only devote so many hours per week to their jobs, perhaps with promotions or bonuses stretching their income upwards, but only so far. Capital, on the other hand, has no such constraint on how much it can accumulate and grow, with money invested into successful enterprises creating the ability to generate returns effectively without limit. The fact that the proportion of the world's population who are billionaires is such a tiny minority testifies to the potential for capital to concentrate wealth.

Most countries seek growth by attracting capital from private investors to fund, among other things, manufacturing plants, infrastructure upgrades, new technology, housing, commercial property, mining, farming, forestry and fishing. Various strategies are adopted to attract such investment, such as lowering taxes and minimizing regulation. With more money flowing into a growing economy, there has been a tendency for the rich to get richer. Simon Szreter told me how economists looking at this have found a rather strong tendency for returns on capital to exceed the average growth rate of the economy. 'It's the old adage,' he says, 'money makes money.'

If policies to attract investment are implemented at the same time as measures to restrain wage growth or to limit the influence of organized labour unions, then that tendency towards greater wealth concentration and deepening inequality can become even more entrenched, as pay packets are held back while returns on capital suffer fewer restrictions. Other factors that exacerbate this trend include the sale of previously public assets, such as water and power utilities, and deregulation to reduce compliance costs, for example by rolling back on rules to protect the environment, thereby freeing up more money to allocate as profit, rather than upgrading assets to cut carbon or protect rivers.

One example of how this can play out was seen in England in 2023, when an attempt was made to repeal laws requiring house-building companies to pay to clean up the sewage pollution that would inevitably come with the construction of new homes. This led to a jump in the share prices of major house-building companies, as investors reacted to a deregulatory move that would push costs down and profits up. These shares were more attractive because the companies would need to spend less on environmental schemes, thereby leaving more money to pay out to investors as dividends. The government justified this step to weaken environmental law because it would, claimed Prime Minister Rishi Sunak, increase GDP through stimulating economic activity worth around £18 billion. The boost to billionaire wealth arising from an increase in share value coming in the wake of Donald Trump's 2024 election victory, and his very clear deregulatory agenda, including on climate change, presented a dramatic example of the close gearing between diminished ambition to cut pollution and the expansion of extreme wealth. These are but two among a great many examples of similar choices going back decades.

Another pro-growth policy agenda that has run alongside a focus on cutting 'red tape' and regulation is that of reduced taxation. Like other pro-growth policies, implicit in this is the idea that reducing taxes is good for society as a whole and not just for those with higher incomes. The year 1979 was when the UK reached 'peak equality', with many of the free-market economic policies enacted after that point effectively driving a widening gap between the most well-off and the least.

For instance, two major tax cuts introduced by President Reagan reduced top-rate taxes substantially in 1982 and 1987.[12] In the UK, taxes on the rich dropped significantly under the Thatcher administration, with major tax cuts in the year of her first election win, 1979, and then again in 1988.[13] Tax cutting, including for the very wealthy, has been a prominent right-of-centre political theme ever since. In the UK, the brief premiership of Liz Truss in 2022 saw a renewed emphasis on this kind of economic theory. Since the 1980s, many other countries have taken this path, reducing taxes for those on higher incomes.

Part of the response to any criticism of these policies is to make the rather counterintuitive claim that they are in fact pro-poor. The logic goes that wealth trickles down through societies, as the wealthy contribute more to growth in GDP by hiring more workers, paying better wages and investing more. In other words, the idea is that the savings the richest make on their tax bills benefit everyone else. This was one of Adam Smith's key conclusions, and it remains alive and well today, as faith is placed in the 'invisible hand' to fix what might otherwise be downsides that can accompany more freedoms for capital.

The growing inequalities seen in many countries across the world during recent decades reveal, however, the limited extent to which this dynamic is real. Most of the wealth created via more growth is captured by the richest, thereby suggesting that policies to enable them to keep more of it are unlikely to flow down by economic gravity alone. For example, in 2020, researchers at the London School of Economics reviewed the impact of tax cuts on the distribution of economic benefits across a range of major economies going back decades. They found that tax cuts aimed at stimulating growth tended to increase the income share of the richest one per cent.[14] At the same time, this research found no discernible positive impact on economic growth or unemployment, with GDP per capita and unemployment rates of countries that cut taxes on the rich and those that did not remaining nearly identical after five years.

The geographer and academic Susan Smith points out that 'the shift in the 80s towards the notion that the best way to achieve growth is to free up markets has meant that markets tend to be unequal. We were urged not to worry about that because for a little bit of inequality we could get lots more growth and that would trickle down so that everyone wins.' Having spent decades looking at the literature on this subject, she is not convinced. 'I don't subscribe to it because the empirical evidence doesn't support it [...] wealth has trickled up over time, to the top one per cent and ten per cent who have accrued an increasing proportion of the growth that we've seen.'

Having said this, it is important not to lose sight of the extent to which growth has helped reduce overall poverty. Smith acknowledges that fact, but she also notes that it is a limited phenomenon which relates to the first stages of growth. 'Actual real investment and maybe trickle down is part of people coming out of poverty in some places, but it isn't the dominant picture if you look at the whole.'

The fact is that the benefits of growth have been unevenly distributed while at the same time economic expansion has placed an increasingly heavy strain on natural systems. These are major issues for all the reasons we have seen so far. These include the extent to which less-well-off people tend to suffer the heaviest impacts of pollution, climate change and lack of access to green spaces, as well as the ways in which inequalities have stymied political consensus. Individual countries' environmental progress is slowed down or blocked because of concerns about impacts on the less well-off. And there is simply not enough environmental space to accommodate the demand created by present patterns of economic growth, never mind increased consumption levels driven by pro-GDP growth policies of the kind that are so prevalent now.

Despite our growth-led economic system's tendency to increase inequality and drive environmental damage (while blocking action to stop it), we remain trapped. We are trapped in a mindset that was born in a world that no longer exists, a world in which natural resources were seemingly inexhaustible and technology was assumed to be capable of meeting every challenge. It was, moreover, a mindset that had its origins in the Industrial Revolution and the age of empire, when the approximately one billion people comprising the global population were utterly unaware of planetary environmental change. They did not know that the coal they were putting into factories, locomotives and ships would place the world on a path to existential crisis two centuries later. We have entered a new reality, however: the Anthropocene epoch, in which some of the assumptions that gave rise to our modern world are now dangerously out of date. Yet for most of the people making the biggest decisions, our economic system and the suppositions that underpin it persist unchallenged.

GREEN GROWTH

As the truth has finally dawned about the environmental crisis, questions have sometimes been asked as to how GDP growth can continue indefinitely on a finite planet, especially when critical ecosystem limits are already being exceeded. This has led to discussions about 'green growth' and how it might be possible to decouple the expansion of GDP from environmental damage. There have also been various commitments in global treaties and national policies in multiple countries to implementing 'sustainable growth' and aligning economic goals with social and ecological aims. More recently, in the UK there was a policy emphasis on 'levelling up', an idea to harness economic growth, but in ways that lead to those who have hitherto not benefited from development getting more of its upsides.

All these ideas, however, whether linked with equality or with environmental questions, still depend on GDP growth, and when push comes to shove it tends to be 'growth' that is pushed forward while the 'green' gets shoved out of the way. This is especially true during times of economic hardship, when leaders are more willing than ever to ditch environmental commitments. The fact that we inhabit a world of rising inequality at the same time as environmental targets fail to be met really sums up how commitments for green, sustainable and fair growth have yet to be translated into practical positive effects.

Tim Jackson has seen how, when the pro-growth agenda is confronted with the prospect of finite resources, the focus shifts to technological solutions. 'That will give us this future where we have less impact on the planet and keep growing,' he says. 'This idea of green growth is very attractive and seductive because it offers people exactly what they want. Yes, you can keep growth. Yes, you don't have to worry about growth dependency. Yes, we keep getting better and better. Yes, we sell more stuff. And hey, we fix the environmental problems through technology.'

He is, however, sceptical as to how this line of thinking pans out in the real world. 'You find historically that we haven't really done a very good job at that. We've done some relative decoupling, we've made things a little bit more efficient, but the overall expansion has outweighed all those improvements. And we're still trashing the environment and have no hope of staying within our planetary boundaries.' He recognizes calls for more investment to get on the right track but points out that this is not happening. 'The scale at which you'd have to introduce technology to make GDP workable environmentally is of heroic historical proportions, something that's not ever been done, and there's not much indication that it's happening. So maybe you shouldn't bank on it to save the planet.'

Part of the problem is the extent to which our present economic system has become so fundamentally embedded, with deep roots that dive down through history. Founded on the exploitation of natural resources and expanding consumption to generate returns for the enterprises involved, it has lifted living conditions for billions of people, but at the same time tended to degrade the environment and concentrate wealth. Our modern economic system has been in gestation for centuries, starting with the early days of empire and industrialization. It has swept all before it – socialism, localism and Indigenous societies. Alternatives to the now nearly universal approach to how markets and capital are harnessed to drive growth in GDP are rarely entertained.

When it comes to avoiding ecological disaster, how the idea of 'green growth' will work is, to put it mildly, not yet clear. The simple fact is that when it comes to more GDP growth of the present kind, there is simply not enough planet to accommodate everyone having lifestyles like most better-off people presently have. This is of course exactly the intention of growth and the strategies that lie at the heart of it – to spread higher levels of consumption, albeit unevenly, and with rising environmental impact a consequence.

The promise of better lives is one reason why growth remains so dominant, as citizens aspire for more of what they see the better-off enjoying. Indeed, this aspiration is essential for generating the political and popular consent needed for the present economic system to continue. Henry Wallich, Yale professor and governor of the US Federal Reserve during the 1970s and '80s, believed that growth and the hope it created made income differences tolerable. He went so far as to say that 'growth is a substitute for equality of income. So long as there is growth there is hope, and that makes large income differentials tolerable.'[15] In other words, if there is a plausible prospect of a better life, most people go along with inequality. Hence growth is not only an economic necessity, in a highly divided world it is also a political one.

No wonder it has become so central an organizing concept. The advocates and beneficiaries of growth, especially those promoting the neoliberal version, have worked relentlessly via think tanks, lobby groups, university economics departments, business schools, broadcasters and the columns of major newspapers to embed their ideals of lightly regulated free markets into policy and practice and into the very philosophy of business leaders and investors, the manifestoes of political parties and the popular consciousness of the public to the point today where there is no real alternative being discussed. We must find an alternative, however, and urgently, if we are to survive, never mind thrive, in the Anthropocene.

Those seeking more GDP growth of course have a different perspective. They see the triumph of a system that has stood the test of time, resolutely resisting the reality that it now faces its ultimate test, as rapid global heating, pollution, mass extinction and the depletion of vital natural resources ramp up. Despite talk of green and sustainable growth, all these things are getting worse.

In responding to this inescapable fact, it has been tempting for some to divert attention away from the economic system and instead place the focus on population growth and how since 1950 the number of people in the world has about tripled. What is generally not mentioned in this frame of reference, however, is that during the same period, the size of the global economy has grown about tenfold, meaning that, on average, per capita wealth has gone up more than threefold, leading to a huge average increase in how much each of us uses up.[16] And what we all use up is, of course, a central element in the modern system and in the continuing drive for more economic growth.

That quest for economic expansion via the extraction, processing and sale of more and more stuff is the driving force behind the businesses that inhabit those impressive glass, steel and concrete edifices that give the City of London its modern character. It has a name: consumerism – and, in common with other aspects of the environmental conundrum, it is entwined with questions of social inequality. Let's have a look at it next.

10

Consumerism, Status and Trust

The engine of economic growth is tuned, revved and accelerated to expand GDP. Into its tank pour ever-growing quantities of the high-octane fuel of mass consumerism, and from space some of the exhaust emissions can be seen. Satellites far outside the Earth's atmosphere have captured images of a vast expanse of discarded clothing sprawling across the Atacama Desert in northern Chile, covering several square kilometres of the arid landscape with unwanted apparel.[1]

This enormous heap of branded jackets, sparkly shoes, T-shirts and shorts is a powerful symbol of our global economic system. Clothes are manufactured in China and Bangladesh and shipped to the USA, Europe and Asia, and then, if no one buys them, they are dumped in the Atacama Desert. Waste is baked into the consumerist business model on a dizzyingly vast scale. In one of the driest places on Earth, Christmas jumpers and ski boots are among the millions of items that contribute to a mound of discarded consumer detritus growing at a rate of about forty thousand tonnes per year.

We can thank the rise of 'fast fashion' for this. The mass production of poor-quality clothing which enables consumers to keep pace with rapidly changing trends at low cost has proved a successful strategy for generating growth and profit for manufacturers and retailers. But, like other mass-market consumer-facing sectors, it comes with a hefty environmental price tag. When I began my environmental work in the 1980s, clothing was never really a major issue. It was energy, transport, pollution from industrial processes, aspects of agriculture, timber production, industrial fishing and built development that were causing the most damage to nature.

The fashion industry is now a highly polluting industry, however, responsible for about a fifth of the wastewater produced across the world. It also emits more greenhouse gases than all international flights and maritime shipping combined.[2] Much fast fashion is non-biodegradable, made from synthetic fabrics manufactured from oil. Microfibres are building up in the sea as our washing machines discharge tiny fragments of clothing into wastewater. About a third of all microplastics now drifting in the ocean come from the laundering of synthetic fibres. They travel through sewage-treatment works, down rivers and into the marine environment, where they pollute food webs. And clothing left in the open, or which is buried, contaminates land, with the synthetic fibres, like the plastic in the ocean, taking centuries to degrade.

The fast-fashion industry has grown at a spectacular pace. Clothing production doubled between 2000 and 2014 (to put that into perspective, the population increase was only about 15 per cent).[3] Behind that remarkable statistic lies an accelerating cultural shift which means that consumers are offered constantly changing collections available at low prices. Advertising and marketing encourage this frequent purchasing and discarding of clothes. On its own terms it has been a winning strategy. According to the McKinsey 2019 'state of fashion' report, the average number of clothing items bought by consumers rose by 60 per cent compared with the situation 15 years before, while at the same time consumers ended up keeping what they bought for only half as long as they used to.[4] And the relative price of clothing has also plummeted, with the proportion of household income spent on clothes in Britain dropping between 1957 and 2017 from ten to five per cent.[5] Given the current impact of the sector, it cannot meet global environmental goals and still keep growing. Indeed, should the sector carry on with business as usual, it could see its greenhouse gas emissions expand by 50 per cent by 2030, which is basically the opposite of the 50 per cent cut needed by 2030 (compared with 2017) to remain on track to meet agreed global goals.[6]

The fast fashion that is so firmly embedded in our culture contrasts sharply with the world I knew as a child. I have vivid memories from the 1960s and early 1970s of being taken to Shepherd & Woodward on the High Street in Oxford to be bought clothes. Above all else, my mother was concerned with quality. We didn't have a great deal of money, but she would nonetheless pay more for items which would last, as well as look nice. The notion of clothing being disposable was so alien an idea that it would never have entered her mind.

During those times, consumerism was on the rise, but it had not yet taken on the extreme dimensions that we see today. And, of course, it's not only an explosion in the amount of clothing being consumed that is the problem. It's also the huge increase in electronic gadgets and disposable consumer goods, from phones to vapes to throwaway cups to cheap furniture and the vast number of cars manufactured each year, which all sooner or later become waste, like that vast pile of clothes in the desert, much of which was never even worn.

Such is the scale of demand for the resources needed to manufacture all of that, that the point has been reached whereby between 2016 and 2021 the world consumed a quantity of raw material nearly equivalent to all the resources used during the entire twentieth century.[7] The price paid in climate-changing emissions, pollution, loss of nature and depletion of resources is huge, and increasingly unaffordable.

THE RISE OF CONSUMERISM

Like everyone else who grew up during the post-war years, I witnessed the explosive rise of the consumerist culture – one of the most striking changes to have swept through western societies. It was not a new phenomenon, however, even during the 1960s. The origins of unnecessary consumption date to the nineteenth century, when the rise of industrial production required market demand to keep up with the productive capacity of the factories and their new technologies. Spending on unnecessary material goods associated with desire and lifestyle was actively encouraged, so that those investing in the expansion of manufacturing could secure a return on the money they'd bet on the success of particular enterprises.

Sales promotion became ever more important, with enticing messages on the virtues of new tableware or bedlinens making an ever-greater impression on consumers. So explosive was the rise in demand that in the USA the production of consumer goods expanded twelvefold between 1860 and 1920, a period during which the national population about tripled, thereby suggesting a roughly fourfold increase in per capita consumption.[8]

Key to driving demand was the propulsive power of envy, unleashing people's previously dormant acquisitive instincts. Opulence expanded beyond the estates of the wealthy and into the homes of city workers. Shops displayed a growing array of attractive goods to more and more people. Consumers showed off their new purchases and helped to foster a sense of fashion and desirability – harnessing the power of 'I want'.

The 1890s saw the arrival of the world's first department store and the earliest mail-order shopping. As consumerism gained momentum, there was a shift from the sale of products with self-evident utility to mostly non-essential goods. Business models evolved to keep pace with the shifting culture and the opportunities it presented. The Hudson's Bay Company, founded via a royal charter in 1670 to exploit the natural resources of British North America, again presents a case in point.[9] During a visit to Canada in 2022, I was interested to learn that this early entity of empire continues to trade today, including in the form of swish downtown department stores located in numerous Canadian cities, and called The Bay.

I went inside the Montreal branch to have a look. Located in an impressive four-storey building constructed during the late nineteenth century from imported Scottish sandstone, it dominates the busy shopping district, where it occupies a prime corner location. Stomping snow from my boots as I went from the freezing street into the warm comfort of a modern consumer paradise, I was met by a wall of beautiful twenty-first-century products and brands. The Bay promotes itself as 'Canada's Life & Style Platform', presenting a vast array of desirable goods ranging from furniture to perfume and from kitchenware to electronics. In a nod back to the origins of the company, customers can buy a modern version of the canoes used by the fur hunters of centuries gone by. The Bay is today the Hudson's Bay Company's flagship brand, and with a history going back more than 350 years, it is the oldest company in North America. Not only can customers visit the shops, they can also buy through internet shopping – the modern equivalent of mail order.

The Bay sums up the transition that took place over a few centuries, whereby gradually capital not only sought returns from exploiting natural resources and from industrialization, but it also began to deploy consumerism as a core economic strategy. As the beaver furs ran out, the company diversified its investments into land and retail. But it wasn't only about technology, logistics and business models; there was a great deal of psychology involved too. The entrapment of furry mammals was replaced by an apparatus that set out to entice consumers, influencing cultural norms so that the acquisition of more goods could become equated with happiness and achievement, and using the cult of the new and the democratization of desire to power whole new dimensions of growth. Where human instincts had once been harnessed to meet need, they were increasingly being tweaked to unleash greed.

MAD MEN

While wealth doesn't trickle down very quickly, desire and fashion certainly do. The psychology of how ideas spread has led to the rise of one of the world's main creative industries: advertising. The TV series *Mad Men* charted the story of fictional advertising executive

Don Draper during the 1950s and '60s. It was a window into the world of smart people, glamour and big money brought about by the rapid growth of consumerism. Even by the 1950s, though, the practices of framing, image making, brand creation, sloganeering and message promotion were well advanced, and had been for some time.

One of the first writers to document the cultural change that took place at the turn of the nineteenth century was the American economist and sociologist Thorstein Veblen. His description of conspicuous consumption was set out in his 1899 book, *The Theory of the Leisure Class*, which defined and examined the practice of buying goods primarily to give an impression of status and wealth, rather than for practical or aesthetic reasons.

Veblen grew up in a world driven by need rather than luxury, but he would witness during his own lifetime a period of social change characterized by the accumulation of huge wealth among a new class of industrial and finance tycoons. These super-wealthy bankers and industry titans were not aristocrats or landed gentry, so they sought to signal their social standing through other means. They went yachting, played polo, bought huge houses and travelled, and in the process showed off their success. But the falling prices of some luxury goods, such as watches, made them more available to the masses, meaning that these newly rich people had to constantly find new ways to display their status to stay ahead of others who might be trying to keep up.

Veblen came up with the concept of 'pecuniary emulation', which was basically a way of describing wasteful and extravagant consumption that serves no other purpose than to impress others and instil envy. Consumerism thus exploited basic human psychology to create a dynamic that led to an unquenchable demand for resources, not to meet need, but in an effort to keep up with the income group above. Along with this culture of wasteful consumption came the myth of the self-made individual and the assumption that the poor were to blame for their poverty. Wealth, or the appearance of it, increasingly came to be seen as a sign of good character and achievement.

The rise of this new culture presented more and more business opportunities, and thus drove research into how to connect consumers with products. The American writer and theorist Edward Bernays worked at the cutting edge of this shift. Remembered as 'the father of public relations', he outlined how an organization could carefully manage the information about itself that it released into the world.[10] He worked out the power of thoughtfully honed messages that were not only about facts but also about psychology, and helped create tools for those seeking to influence the public via advertising.

Bernays's ideas were shaped by his experiences working for the Committee on Public Information during the First World War. It was an official US agency created to influence public opinion at a time of national crisis and to garner support for the war programme, particularly on the home front. He realized that 'what could be done for a nation at war could be done for organizations and people in a nation at peace'.[11]

He is credited with making smoking in public socially acceptable and encouraging many more women to smoke. He successfully urged Americans to eat bacon and eggs for breakfast and promoted bananas as a popular healthy snack.[12,13] His methods shaped the content of media coverage, persuading newspapers and magazines to publish medical opinions about products and getting famous actresses to smoke where they would be photographed, along with other kinds of product placement. He was Sigmund Freud's nephew and realized how his uncle's pioneering work in psychotherapy could be harnessed to promote retail therapy. Bernays is seen as the founder of consumerism, his ideas based on an understanding of what makes humans tick. He realized that by delivering carefully selected images, messages, frames, role models and partial information to consumers, they could be convinced that they could transform themselves through shopping.

Professor Sean Nixon is based at the Department of Sociology at the University of Essex in Colchester, England. He has devoted a great deal of time to understanding advertising and its history from when, in the late nineteenth and early twentieth centuries, a recognizable industry began to emerge.[14] The early advertising industry started in the USA and spread to Europe, and Nixon has charted how it was linked with a shift from local markets to national ones backed by national brands. He told me how it began with the press. 'It's national and big city newspapers and the alliance between the new advertising agencies and the press that drives this new model. They hire agencies to help them grow their revenue and then over time they also start producing the adverts for them.' As the press grew, more advertising was needed to cover their expanding costs. In the circulation wars that followed, it was the newspapers most attractive to the advertisers that made the most money for their media mogul owners.

The rise of advertising was accompanied by other major developments, including that of mail order, which enabled markets to operate across the vast North American continent. This in turn was facilitated by the rise of the railways. Consumers were offered what Nixon describes as a 'cornucopia of consumer delights' via publications like the Sears catalog. First published in 1888, it became a benchmark for what Americans thought they should have in their homes, driving higher and higher levels of consumption. It also became a major driver of economic growth; as one market became saturated, manufacturers and advertisers looked for new opportunities. 'You have to create a new dissatisfaction to create demand for something else. There's a kind of treadmill that goes ever up, with advertising promoting ever higher levels of consumption,' says Nixon.

He found that the bulk of advertising during the first two thirds of the twentieth century in Europe and North America was focused on households. 'There was a big obsession with health, hygiene and cleanliness.' This was used as an entry point that he says ultimately came to shape cultural norms. 'Advertisers saw themselves as playing educational roles in bringing new standards for hygiene and cleanliness, but they're also setting new norms about the levels of consumption that should be acceptable, what you should have, how much you should be spending. They promoted to a mass of the population higher levels of expenditure around what constitutes the kind of normal patterns of consumption to the point where you've got to have a lot more stuff in your home rather than just basic things.'

Nixon's research found that during the interwar years advertising began to move beyond genuine human needs for food, shelter and sociability. Instead, it now revolved around new ideas about comfort, cleanliness and status. 'The trick was to create a series of problems for consumers so that they can then bring in the product as a solution. Often these are new needs that aren't really what you might call basic needs. The selling of motor cars is the obvious one. It clearly has practical uses for mobility, but advertisers link it very much with status and the need to have the latest model.' This focus on novelty was another key theme for advertisers. 'New is good and you need the newest version. You can see that now with the technology sector where a phone that's six months old is seen as out of date.' He also highlighted the case of clothing. 'Fast fashion is now just pushing this incredibly accelerated cycle of change, where you just throw away clothes that have been worn a couple of times. Because they're cheap, that creates a sense of not valuing them, not valuing the work in the clothing, not valuing it as a piece you want to repair and look after. It's definitely pushing a culture of waste on the basis of giving access to the latest thing, the newest thing.'

Today, product promotion underpins one of the world's largest economic sectors. In 2022 the size of the global advertising industry reached about US$700 billion.[15] It attracts such vast investment, with the best creative brains being hired and media space being bought on TV, billboards, magazines and digital media, because it works. Global consumer companies are willing to spend fortunes to harness the expertise of the public relations professionals. Coca-Cola is one of the world's most recognized brands, and from 2014 to 2022 spent about four billion dollars per year to maintain its prominent position in the minds of the billions of people who buy its products.[16] With that huge brand reach comes great power, which has the potential to trump messages on public health.

In the UK, the main reason children receive anaesthetic is to remove decaying teeth.[17] The principal cause of this tooth decay is too much sugar in their diets, including from fizzy drinks.[18] Treating the consequences of this costs the NHS tens of millions of pounds every year.[19] Meanwhile, on the other side of the world, mothers in some communities in Mexico have moved their children onto Coca-Cola after breast milk.[20] Coca-cola uses a highly localized strategy, in such areas, with advertising boards made bespoke to particular communities. In one campaign in a poorer rural area with a high proportion of Indigenous people, Coca Cola advertising used images on billboards portraying local traditional dress. The fizzy drink has become so embedded into the social fabric that families reportedly place Coca Cola bottles on family altars.[21]

This is an example of the power of advertising to insert products into the cultural and even spiritual fabric of communities, and in that particular case, in a country where many people don't have access to safe drinking water. Mexico is not alone. Coca-Cola spends fortunes promoting its products in more countries than there are members of the United Nations.[22] The power and reach of its brand is legendary. I've glimpsed that myself, working in some of the remotest areas in the world where, even if there are no roads or modern medical facilities, Coca-Cola can still be found. For instance, while conducting field research with Brazilian scientists back in 1990 in the remote north-east of that country, where there were no roads to speak of, I saw bottles of Coke reaching the remotest communities in crates carried on the backs of mules.

Advertisers use a great many tricks to trigger that critical decision to buy. Digital media uses individuals' data to enable personalized targeting to be carried out. Whatever route these companies take to get our attention, whether on a billboard or a smartphone, an important framing theme concerns the psychology of belonging and the very human propensity to judge our own status through comparisons with others. Are we keeping up or are we falling behind in the race for success, status and respect? Posing that question via the sometimes subliminal messages embedded in advertisements can fuel a powerful psychological phenomenon, and it is one that is more pronounced in less equal societies.

STATUS ANXIETY

Status anxiety describes the internal discomfort felt by many people when perceiving themselves to have lower social rank than others. Such anxiety is more prevalent in societies where inequality is more pronounced and obvious. Research has also linked status anxiety with another phenomenon known as status consumption. This is a response to a motivation not to feel, or be seen to be, left behind, sidelined or less successful; people often purchase status-laden products, such as expensive luxury brands, as a way of combatting the stress that can accompany such feelings.

Researchers at the Mind, Brain and Behaviour Research Centre at the University of Granada in Spain have delved into the relationships between status consumption and economic inequality. They found that the greater the income gap, the more people were inclined to engage in status-seeking behaviour, with participants in one study reporting higher status anxiety when differences were more pronounced.[23] When conditions were less equal, participants were also more materialistic and more relaxed about indebtedness.

While there is still work to do in untangling the complex relationships between inequality, psychology and consumption, there is growing evidence to support the hypothesis that perceived economic inequality has a major effect on status consumption, status anxiety, materialism and attitudes towards debt. Individuals adapt to the social pressures of inequality by seeking out, acquiring and displaying goods that they believe will convey positive signals to others regarding their social status.[24,25] The more unequal the society, the more powerful the effect becomes.

Sean Nixon told me how his work revealed the extent to which advertisers 'became very good at picking up on anxieties and then finding a way of linking those to products, telling a story about products around those anxieties.' One enduring tactic used by advertisers in this respect was linked with status. 'In Britain it was about keeping up with the Joneses. If your neighbours had the car on the front drive, you ought to aspire to something similar. Being left behind socially and symbolically was an anxiety that was played upon. Envy and social aspiration were built into lots of TV advertising. Even at the simple level of soap powders, adverts would ask if your sheets were whiter than Mrs Jones's, with shame coming from having sheets that were not white enough.'

People living in societies with bigger income differences spend more on status goods and are more likely to experience debt and bankruptcy.[26] And this is not only about the less well-off trying to keep up; the rich also vie with one another, leading to situations where in less equal societies all income groups feel more anxiety about their status. Even billionaires experience jet and yacht envy, their purchase of astronomically expensive status goods driven in part by a desire to keep up with each other and stay on top. The modern *Miss Conduct* superyacht that sits in the docks at Bristol, adorned with a helicopter, was not bought for utility.

The top end of the market for status goods, such as expensive watches, fragrances, leather goods, clothing and eyewear, is huge, worth about US$354 billion in 2023 (a number which doesn't even include luxury cars, private planes and yachts).[27] As might be expected, the countries with the largest economies (and greatest levels of inequality) account for the biggest shares of sales. The USA's revenue in the luxury goods market amounted to more than US$75 billion in 2023, making it the largest luxury goods market in the world, accounting for 21 per cent of the total, while having just over four per cent of the global population. But luxury is increasingly, of course, not only a western obsession.

As economies have grown (and grown more unequal), so demand for luxury has expanded in the emerging economies too. China's revenue in the personal luxury goods market totalled more than US$53 billion in 2023, accounting for about 15 per cent of the total. Even with demand being hit hard by COVID-19, Hong Kong, with just 0.09 per cent of the global population, still saw some four per cent of the total number of luxury goods purchased being bought there.[28,29] In the USA, the largest segment for luxury sales was fashion, while in China (including Hong Kong) it was luxury watches and jewellery. Interestingly, spending on advertising tends to form a larger percentage of GDP in countries with greater income inequality.[30] Perhaps it is less beneficial for companies to invest in such promotion in more equal societies.

In any event, the sophisticated promotion of status goods helps drive the ever-steeper upward spiral of consumerism. Among the consequences is the astronomical demand for natural resources, which is set to go far higher should present trends continue. The effects of consumerism are not only environmental, they are socially corrosive.[31,32] Those who can't keep up are excluded, causing a level of social stratification on top of what was already there to begin with. This can be made worse still through the effects of debt, as household resources are diverted away from social goods towards consumer goods, thereby exacerbating inequality even further.

Lying behind all of this has been the rise of the advertising industry and the shaping of the rapacious consumer culture that has now spread to all corners of the world, and which still grows and spreads its impacts on our minds and choices. Having begun with newspapers and mail-order catalogues, consumerism expanded with billboards and posters, then raised its reach via radio and TV, and is now supercharged via digital media, round-the-clock internet shopping and the power of influencers. Using short video formats, brands increasingly hook up with prominent individuals to gain access to their followers. These are in turn run by algorithms that harvest and process data while seeking to keep people fixated on their screens for as long as possible. Nixon says this tactic hinges on ideas of authenticity. 'They're mostly around lifestyle and consumer goods and appear to the viewer not like advertising because you feel like you know the influencer or follow them, you're a fan. I worry the influencer industry has taken consumerism up another level at the time when it needs to be going down a few levels.' Pressure on the biosphere that sustains life is thus under growing pressure from an electronic 'buyosphere', geared to maximizing sales of what are largely unnecessary products.

That digital sphere is fuelled by some of the world's largest corporations, who all have an interest in its growth. At the time of writing, the four largest such companies, including Apple, Microsoft, and Nvidia, already have a combined value of nearly twelve trillion dollars, with Apple alone having a greater value than the annual GDP of Australia and South Korea.[33,34] The business models operating in this new frontier of consumerism harvest vast quantities of personal data, recording preferences and eye time on their platforms and devices. As artificial intelligence becomes more powerful, this huge economic sector is at the cusp of reaching a new level of impact to sell us yet more.

As sales methods have become more and more sophisticated and omnipotent, and as the environmental pressures have grown, there has been very limited counter-messaging to reduce overall consumption, with product manufacturers instead tending to harness the relationships between social status and consumer choice to promote what are portrayed as lower-impact status symbols, such as electric vehicles.

An A-list Hollywood star driving the latest electric luxury car might send a slightly different signal about the possibility of aspiring to a cleaner and greener lifestyle that is also high status, but as we saw in Chapter 7, there is simply not enough stuff for everyone to consume in the way that celebrities, or the western middle classes, do, even when what they do is arguably a bit 'greener'.[35] With that in mind, it is hard to see how expanding inequality-driven consumption can be compatible with a credible strategy to meet the demanding targets that must be pursued to protect the integrity of our planet's life-support systems. I have found, however, that this is a very hard point to get across.

During the 1990s, we tried various ideas at Friends of the Earth to challenge consumerist culture. This included working with campaigners who came up with the idea of 'No Shop Day'. An actual 'No Shop' was opened in London. It had nothing in it, and when shoppers entered – and a few did – instead of being sold products, they were engaged in conversation about alternatives to overconsumption. The idea of No Shop Day was to convey messages about sufficiency and having enough for a good life, rather than endlessly looking for more.

Later on, 'Buy Nothing Day' came to the UK from North America, where it had been introduced as a counterpoint to Black Friday. Held in late November, Buy Nothing Day was the brainchild of campaign group Adbusters. This activist network was established in Canada in 1989 to challenge consumerism. Among its methods were attention-grabbing creative visuals that sought to open the eyes of those who it believed were being manipulated by advertising.

I loved working with Adbusters, with their mischievous counter-cultural approach. Among the tactics they employed was a practice called subvertising, whereby campaigners would change, spoof, satirize and generally undermine adverts to say what the advert wasn't saying, with the aim of getting consumers thinking. One example was a roadside poster with a picture of a British Airways passenger jet with a message saying, 'We are turning business class green,' adding cheekily, 'with the world's first on-board golf course'. But for such grassroots campaigns setting out to challenge the power of the advertisers and the consumerist culture that they had spawned, this proved hard work.

We found it hugely difficult to have serious debates about the issues in play, with, for example, many radio and TV interviewers pushing back hard with questions which avoided key concerns about environmental impact. 'So you don't want people to have nice things?' 'Would you like everyone to stay at home and never go shopping again?' 'Why are you against economic growth and development?' 'Why should rich-country campaigners stop poorer people enjoying what we take for granted?' Campaigners were portrayed as trying to stop people enjoying themselves and as being against the interests of families working hard to improve their lives.

I asked Nixon why it is so hard to challenge the consumerist status quo head-on, even in the face of a global ecological emergency. 'It's like the old hair shirt, a kind of Puritan view of suffering for your cause,' he observed. 'That's hard to sell as a positive message to people that want to have a nice life. There is an issue about trying to get people away from this relentless cycle.' Such difficulties highlight the enormous cultural momentum that sustains the current situation, with those communications challenges one reason why in recent years there's been relatively little emphasis from campaigners and environmental advocates on buying less stuff.

'Going for greener' has been an easier message to communicate than 'going for less'. That, coupled with a strong emphasis on recycling, has been the preferred strategy for many campaign groups, leaving the underlying situation largely unchallenged. Predictably, this has led to numerous accusations of 'greenwashing', whereby companies and the advertisers who promote their products and services have been inclined to overstate the green credentials of what is on offer to consumers.

Irrespective of the physical and material consequences of different product choices, and regardless of whether lower-consumption messages can ever be made to work better than 'green consumer' ones, some very important psychology is in play. This can in different ways distract from environmental and sustainability priorities. In their famous book *The Spirit Level*, Kate Pickett and Richard Wilkinson, social epidemiologists from the University of York, set out research findings which they argued revealed 'why equality is better for everyone'. They drew together data sets from across the world to plot the differences between countries on the basis of levels of inequality, and reached some important conclusions.

I spoke to them to find out what they thought about the relationship between inequality and consumerism. Wilkinson told me that their research confirmed that two things happen. 'We become more worried about how we're seen and judged, so more conscious of status, and when that happens people become more narcissistic,' he explained. 'Studies actually show that if you live in a more unequal area, you are more likely to spend money on a luxury car that looks good and more likely to buy clothes with the right designer labels and so on.' For individuals, he said, there are psychological consequences arising from this. 'People can feel worried all the time about what people think of them, how they are judged. You might start feeling lack of confidence and low self-esteem. You start feeling social contacts are too stressful. You withdraw from social life and can become depressed as well.' These and other insights have led him and Pickett to conclude that social class and social status are not just proxies for risks that are linked with differences in material living standards, but that inequality is a risk itself.

Questions connected to the affordability of energy, water and food quite correctly cite inequality as a cause of material deprivation. What has been less obvious, however, is that inequality also has a corrosive effect on psychological well-being which impacts our social fabric, and thus the ability of societies to deal with collective challenges. Pickett explained how chronic stress linked with inequality contributes to a range of negative social indicators, including increasing levels of violence, that are triggered by people feeling disrespected and looked down upon.

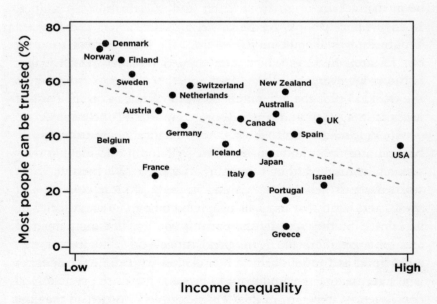

Figure 12

This is just one indicator of how the rather hard-to-pin-down, but no less vital, assets of trust and social cohesion can be eroded by inequality. Wilkinson and Pickett are among a number of academics who present evidence to suggest that it is not consumption that most contributes to contentment, but the quality of human relationships. Wilkinson said, 'If you look at papers on happiness, it's about friendship and involvement in community life, and you can show so easily that that is exactly what inequality destroys.' He also said that 'Inequality undermines people's social connectedness, and that itself is a barrier to achieving policy change to deal with the climate emergency or anything else'.

SOCIAL CAPITAL

The psychology of consumerism is fuelled by inequality, with consumers tied to a system in which an increasing amount of stuff being used up is counted as positive, both because it is culturally rewarded (people with a lot of gear seem to be successful) and because it contributes to growth in GDP. Part of the psychology of advertising is to encourage consumers to believe that their happiness will be enhanced by clothes, cars, carpets, phones and all the rest of the products that carry ever faster-changing fashion values.

When people 'succeed' and keep up in the status race, it is not necessarily a passport to happiness, even among the rich. Social psychologists have been aware of this fact for some time, but it became more prominent when it was given a now widely known name: affluenza. Although the term originated with John de Graaf and colleagues, who wrote a book on the subject in 2001, it was made more famous by journalist Oliver James in 2007 via the title of his book exploring why status consumption was making people all over the world depressed. 'Affluenza' is now included in dictionaries and defined as 'the guilt or lack of motivation experienced by people who have made large amounts of money'. One of the conclusions that James, Wilkinson, Pickett and others have reached is that the misplaced pursuit of contentment via consumerism is damaging not only the environment, but also individuals and society. So why do we continue to accept it, and even demand more?

There is a lot of complexity behind all of this, of course, with researchers delving beyond the correlations between inequality and status-seeking consumerism to try to understand whether the relationship is a causal one. For Wilkinson, Pickett and others, the link is indeed clear and causal, with inequality driving the unnecessary purchase of status goods. This means that inequality increases the environmental impacts that arise from more demand for more resources, but also contributes to the simultaneous erosion of the social capital needed to foster the spirit of collective endeavour that is essential for meeting environmental challenges.

Pickett pointed out that in unequal societies the sense of shared agency and agenda is diminished. 'Social capital and trust are undermined and the idea that we're all in it together dissipates.' On this link between social capital and progress on environmental challenges, research has found that more equal societies recycle a higher proportion of their waste, suggesting that in such contexts individuals are less self-centred and more willing to act for the greater good. This effect is also visible among businesses, with one international survey finding business leaders in more equal countries rating environmental agreements as much more important than those based in less equal countries.[36]

The erosion of social capital in less equal societies has other implications beyond the management of environmental matters. There is, for example, evidence that increasing top incomes over time leads to less political appetite for taxation, which in turn can hit spending on public goods, such as education and high-quality green spaces.[37] Diminished funding for these essentials increases the difference between more and less desirable neighbourhoods, where good schools and attractive natural areas remain in the former, but not in the latter. Inequalities become more entrenched as families try to keep up, with less favoured areas declining further as people with money and opportunity move out to where there are better schools and a nicer environment. As private affluence moves out, leaving deepening public squalor behind, so incinerators, polluting industry and degraded green space are more likely to move in.

There are also implications for democratic politics, with recent research suggesting a link between higher inequality and not only diminished social trust (that is, trust in other people) but also diminished political trust (that is, trust in political institutions and organizations).[38] The way Donald Trump built support for his 2016 presidential campaign deliberately fostered distrust in political institutions with slogans that included 'drain the swamp' and 'lock her up' (referring to his election opponent, Hillary Clinton), highlighting the extent to which sources of anger, including those linked with social inequalities (and the USA is one of the most unequal western societies), can be harnessed to create electoral advantage. A similar strategy was deployed with dramatic effect in 2024, when Donald Trump once more won the Presidency. Politicians who use these kinds of framings often also have anti-environmental policies as a prominent part of their offer. Trump certainly does. The election of Reform party leader Nigel Farage to the British Parliament in 2024 had a similar mix of messages. His party has denied the very existence of human impacts on the climate, and yet this privately educated and wealthy individual, who, before his involvement in politics, worked as a commodity trader, succeeded in being elected to represent the most socially deprived local authority area in England, in part thanks to a platform that included opposition to plans to reduce the UK's climate-changing emissions to net zero by 2050.[39]

As inequality grows and social divisions deepen, wealthy elites can physically separate themselves from the rest of society and its environmental and social ills, while exerting influence through exploiting this declining trust. Through their ownership of media, as well as via political connections and support from well-funded think tanks, they have the means to deploy campaigns with huge ramifications. Academics at University College London writing about how social inequalities impact on environmental issues present the example of wealthy political actors in the USA promoting the belief that the government takes from the 'hard-working' white working class to give handouts to the 'undeserving' poor, immigrants and people of colour.[40] This framing weakens the bonds of social solidarity, making it harder to deal with collective threats such as those posed by climate change, and diminishing citizens' willingness to make changes for others' benefit.

The evaporation of trust can fuel the kinds of populist politics that tend to dismiss environmental challenges as fake, woke or automatically an agenda of the hard left. Their message is to trust *me*, because you can't trust institutions, experts, the political process or, in the case of Donald Trump, even the courts. They demonize views that are contrary to those being pushed, seek to restrict protest, and threaten legal actions to muzzle critics.

The evaporation of trust can also be manifested in the rise not only of social divisions but also of physical ones. On a visit to South Africa in 2002 to attend negotiations marking the tenth anniversary of the Rio Earth Summit, I was struck by media articles that picked out 'heightened security' as an economic success story. On TV, there were many adverts offering security services, and I saw in the better-off neighbourhoods what this entailed: razor-wire fences, surveillance cameras and sealed-off gated communities dividing the city between rich and poor.

I learned how a great many South Africans were employed in 'guard labour'. Private security firms, the police and the prison service were among those seeking new recruits. Data from around the world reveal how this element of national economies increases as income differences become larger.[41] It is good for GDP growth, as the rich procure fences, guns, cameras, security personnel and safe rooms to protect themselves from the consequences of a divided society, but it is very bad for social cohesion and everyone's quality of life, including that of the rich people who've taken refuge inside the fences and evidently live in fear.

In my own experience, I've seen big changes linked with deepening inequality. I grew up in a council house during the 1960s and '70s in a mixed neighbourhood where there wasn't great social stratification, even between those who owned their homes and those who rented social housing. They were mixed up and looked pretty much the same. Local green spaces were well maintained and had a lot of wildlife, birds, butterflies and more, and so did the gardens. There was a sense of common interest and shared reality. This was reinforced by there being only three TV channels and a group of newspapers that broadly reported stories with what was, compared with now, a stronger commitment to balance and accuracy. News and opinion were clearly separated, and efforts were made to get facts right. There was still plenty of debate, but the information was more reliable than it is now.

I went to local schools, did O and A levels, and then attended Bristol University to read zoology and psychology, thereby fuelling my lifelong interest in animal behaviour and my passion for nature. My time at university was backed by a government grant, no fees were charged nor any debt accumulated, and the experience facilitated social mixing, with young people from many different backgrounds brought together. To this day I remain close friends with many of the friends I made there. It is true that back then there was more overt racism and that the gender divide was bigger, but society was economically more equal. In my estimation, this was why the social fabric felt stronger.

As economic and social inequalities have deepened, however, things have changed. The wealthy super elite are more remote than ever from the rest of society, the least well-off more lacking in hope, and the middle classes (along with everyone else) more stressed. Many commentators speak in the 2020s of Victorian levels of social inequality. And that overall trend towards increasing disparity is one of the main reasons why we are now failing to deal with the existential threats posed to civilization by the damage we're causing to Earth's life-support systems.

This is in large part down to how the dominant growth-led system seeks to create more wealth through more consumption (in the process exacerbating inequality). The pile of clothes in the desert, the advertising campaigns that encourage unnecessary and wasteful consumption, the rising prevalence of mental health problems, global heating and the destruction of ecosystems are thus all connected.

With inequality at centre stage in the ecological crisis, not only is reducing it consistent with moving towards sustainability, it is a precondition for it, and not just for the less well-off, but for all of society. The need to create more fairness in the social realm in order to unlock opportunities in the environmental one is clear. But what can be done?

11

A Just Transition

There are few historical parallels that can compare with the world's unfolding ecological emergency, although it seems to me that the Second World War can. When I was a child, my late mother, who was born in 1928 in the Docklands of East London, told me stories about that time. As the years have gone by, it's struck me that there are similarities between the huge efforts needed now to face the ecological crisis and the way British society reacted to that emergency. Her first trip outside the back-to-back terraces came when she was evacuated along with her younger brother Tom to a dairy farm near St Ives in Cornwall. She was 11 years old when for the first time in her life she saw fields of grass with cows, hills and, most amazing of all, the Atlantic Ocean.

Among her recollections was the tale of a German plane that had crash-landed near where she was staying. Watched over by the members of the Home Guard, children paid a penny each to sit in it. Another time, a German aircraft seeking to escape a British fighter plane dropped a bomb next to the farmhouse, shattering the windows, as she and Tom ran down the stairs to find shelter. There was also the time they nearly drowned in a deep rock pool, as well as her account of walking on clifftops where the primroses were so dense that she could not help but step on them, and the story of going to school on the horse-drawn cart that took milk churns to town.

My mother was evacuated as columns of tanks and fleets of bombers smashed through Europe. In 1940, Britain found itself effectively alone against the tyrannical regime of Nazi Germany. Emergency measures were put in place. These ranged from new civil defence arrangements in cities that were about to be subjected to heavy bombing (including the evacuation of children to remote rural areas), to the urgent upgrading of coastal protections, and from a rapid expansion in the manufacture of military hardware to massive investments in new technologies, including modern aircraft and, later on, computers.

The seismic shifts that were so rapidly put in place were accompanied by a realization that if Britain was to survive, never mind win, the war, effort would need to be evenly shared across society. Children from working families were evacuated, income inequalities were reduced via changes to taxation, essential goods were subsidized, luxuries were taxed, and rationing ensured that food and clothing were available to everyone. Princess Elizabeth appeared in public wearing austerity clothing, underlining the sense of common purpose, collective effort and unity that was so vital for Britain's titanic struggle. She joined the army and trained as a motor mechanic.

The threat posed by ecological degradation is, of course, less visible. It moves more slowly, the 'enemy' is harder to identify and, as a result, the narrative for rapid and costly mobilization is more difficult to articulate. But I am not the only one to see some equivalence between war and climate change. In 2021, none other than the Prince of Wales, now King Charles, warned, on the subject of climate change, that we 'have to put ourselves on what might be called a war-like footing'.[1] History reminds us that this war-like footing would not only involve the military, technology and logistics, it would have to embrace social change on the home front too.

There is, however, another crucial difference between war and the ecological emergency: if we respond in the right way, it could be less about sacrifice and more about opportunity. That opportunity includes creating a healthier environment, increasing our resilience to shocks, generating new jobs and industrial sectors, improving national security and enacting sustainable poverty reduction within the ecological capacities of our finite planet. On global heating, there is good reason to believe that it is still possible to do away with neediness and destitution while also reducing emissions. For example, when the Development Data Group & Climate Change Group at the World Bank looked at the climate implications of ending global poverty, they concluded that it need not eat up the remaining global carbon budget.[2] Yes, the environmental space available to eradicate poverty is limited, and it is made more limited by global top emitters' footprints, but with the right approach the impact of poverty alleviation measures can be managed.[3] This will, however, require some significant shifts.

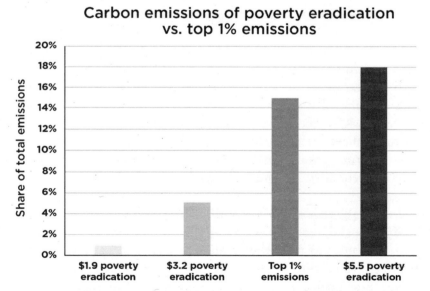

Additional carbon emissions of different poverty alleviation scenarios vs. emission from global top 1% to Bruckner et al. (2022)

Figure 13

I have set out in this book to show how fairness, justice and equality are inextricably linked to the ecological crisis, in terms of its causes as well as our failure to act decisively in dealing with it. Aligning the needs and desires of eight billion people with the capacity of our planet is obviously highly complex, but I believe the priorities can be summed up in just six words: renewable, sustainable, circular, regenerative, restorative and fair.

Renewable as in, for example, making the switch to clean energy. Sustainable as in taking from nature only that which can be naturally replenished and emitting only what can be naturally absorbed. Circular as in mimicking the cycles of nature, over time eliminating waste through smart design, recovering all useful materials and using less stuff to start with. Regenerative sums up the ways we can make natural systems more resilient. Restorative action is needed too, not only to protect what remains of intact ecosystems, but to reverse losses of forests, wetlands, grasslands and more, so they can catch more carbon, replenish freshwater supplies, sustain recovering wildlife, increase the human ability to cope with climatic changes and support health and well-being.

Together, these five concepts would lay the foundations for a secure future. They won't be adopted as quickly or as widely as is needed, however, unless a wider programme is embraced that links them with the sixth concept: fairness. If the programme isn't fair, then it won't work. This is why we need a just transition based on equality, not just new technology and environmental policies.

This will require a plan that embraces steps that might not appear 'environmental', but that will close the gap between official targets and our ability to meet them. There is already a very wide range of international agreements and targets in place, but the problem is that these goals aren't being implemented because governments see social and environmental challenges as separate issues. This is why the most important shifts need to take place within individual countries. Each nation is different, and plans will vary, but I have set out below a ten-point agenda for a developed country such as the United Kingdom.

1. New measures of progress

What is the economy for? The efficient allocation of scarce resources? Improving human well-being? Generating wealth so that everyone's needs can be met? 'To grow' is an answer that makes not a lot of sense, and yet that is how we carry on, assuming that if our economy expands its GDP, everything we want will automatically follow. In reality, what we actually require is an economic system that enables everyone to live fulfilling lives within environmental limits. Such an economy would meet people's needs for shelter, food, water and energy in a resource-efficient manner, while promoting the flourishing of individuals, social cohesion and longevity at the same time as improving the state of nature.

All these ambitions can be measured and combined to create new economic indicators to replace GDP, which only measures our consumption of goods and services. This is not to advocate a 'no growth' policy, but rather to observe that if we changed what it is that we are striving to grow, we could alter the outcomes.

Economic development has transformed human experience for the better in so many ways, but it is important to recognize that the biggest positive impacts of economic growth occur in the early stages of development, when people first make the transition from poverty to a more dignified, secure and comfortable existence. This step occurs at a relatively low level of per capita GDP, and increased consumption beyond that does not necessarily elevate well-being very much, if at all.

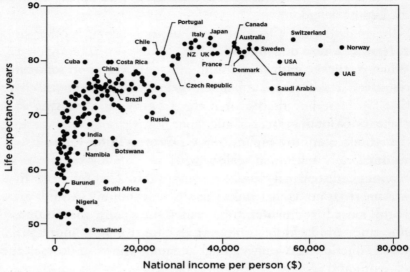

Figure 14

Cuba and Costa Rica have achieved a high life expectancy for their citizens on about a third of the per capita GDP of more developed nations.[4] Cuba achieved developed-country-level longevity despite the scarcity that arose from economic sanctions forcing it to achieve more with less, while Costa Rica did it through a more positive strategy that linked ecological recovery with national development. Between the 1980s and 2020s, it dramatically improved human development indicators while simultaneously doubling the extent of its rainforests. Both these nations are also more equal socially than many richer countries, including the United Kingdom and the USA.

Economist Tim Jackson, whom we met in Chapter 9, does not believe that all growth is of the same quality. 'What's very very clear is that when you go from nothing to around about fifteen to twenty thousand dollars per capita a year you get massive increases in real prosperity,' he said. 'You get better nutrition, you get better health, you get lower infant mortality, you get raised life expectancy, you measure happiness differently. That's what you could call a relationship between the GDP and prosperity in a meaningful sense.'

Beyond that basic level of economic well-being, another story emerges. 'When you look at the shape of the curve after that it tails off very quickly, with diminishing rates of return and sometimes negative rates of return,' explained Jackson, emphasizing that such findings underline the fact that there is not a linear relationship between economic income and improved outcomes: 'there isn't, once basic needs are met, a causal relationship between material consumption and human well-being'.

Once an optimal level of consumption is reached, it is non-material needs that make the most difference. 'This in a sense is the most fundamental thing that's wrong with the GDP as a measure,' says Jackson. 'It doesn't capture that non-linearity, nor the distinction between material activities for which some kind of economic activity is necessary and those aspects which make us human, such as our participation in society and our sense of community and our desire for a clean environment. Those things cannot be captured and yet are critical to our well-being.'

Measures of happiness follow a similar pattern to trends like longevity, with additional wealth beyond a certain level not adding to how content we feel.[5] This too is a very important finding when it is known, for example, that good relationships protect well-being and make as much difference to health as whether a person smokes or not. And yet, even though societies divided by inequality have a weaker community life and suffer from greater stress than more equal ones, we continue to measure growth in GDP as our principal indicator of progress.

There are several alternatives to GDP. They are all based on the observation that for most countries, once economic growth parts company from increased well-being, the logic that justified using GDP is gone. One of these is the Happy Planet Index, which measures which countries are doing best at achieving long, happy and sustainable lives for their citizens. In 2019, the top five countries that best combined high social well-being with a low ecological footprint were Costa Rica, Vanuatu, Colombia, Switzerland and Ecuador.[6]

Another measure, and one that we advocated at Friends of the Earth during the 1990s, is the Index of Sustainable Economic Welfare, or ISEW. This was proposed in 1989 by World Bank economist Herman Daly and philosopher John Cobb as a means of going beyond adding up all the expenditure and consumption taking place across an economy to also embrace environmental and social factors, including a measure of income distribution.

One more alternative to GDP is the Genuine Progress Indicator, or GPI. This calculates progress using measures of personal consumption and expenditure but adjusts them using 24 other components, including income distribution, environmental damage and negative social trends, such as crime rates. It also takes into account the upsides of activities not included in GDP, such as volunteering and domestic work, including childcare, which doesn't appear in national economic calculations, but without which societies would collapse.

One 2013 study applied the GPI to 17 countries from 1950 to 2003 to review their progress on that indicator compared with GDP. It found that while GDP grew massively over that period, the GPI peaked in 1978.[7] The late 1970s marked a time of much greater equality in some countries (the UK included), and it was also about that time when global human demand began to exceed the Earth's biocapacity.[8] The same study concluded that in the countries studied, which contained more than half the world's population, life satisfaction had not improved significantly since 1975.

By placing undue emphasis on GDP while excluding other measures, we end up making choices that lead towards problematic outcomes. Seeking out alternatives is a fundamental aspect of how we must support the changes we need to make.

2. Purpose-led companies
Companies, like the economy more broadly, have also been designed to grow. Growth means shareholders can be paid dividends and investments made in new technologies, businesses and products.

Profit maximization is incentivized via salaries and bonus packages for executives. Successful companies grow, displacing and absorbing competition, sometimes creating vast organizations with global reach. This has served the owners of capital very well and driven the expansion that has enabled the world to develop in the manner that it has, in the process causing environmental destruction and huge inequality. We have reached the point where this model is no longer viable.

Forward-looking companies are adopting a purpose beyond profitability and short-term financial gain for shareholders. Instead, they explicitly seek to create value for society. While making a profit might be a necessary precondition for a company to operate, it is for these types of company not their reason for existing. For example, a water company that aims to generate maximum dividends for investors would have a different strategy to one established to supply clean water for its customers while protecting the environment. Driven by this wider purpose, it would need to make a profit to pay for new reservoirs, sewage works and distribution networks, but profit would be a means to an end, and not an end in itself. This was the direction of travel announced by the new Labour government in the wake of the 2024 general election, with water companies required to place customers and the environment at the heart of their objectives.

When it comes to redefining shareholders and stakeholders, companies might serve the Earth better by integrating on their boards the voices of young people who will inherit the future consequences of business decisions. Countries with Indigenous populations can also embrace the perspectives of those communities on company boards. Māori people are, for example, included as board members in a number of New Zealand-based businesses.

Companies have been setting targets and publishing strategies on environment and social responsibility for decades, but most have remained fully committed to profit growth. As a result, their ecological and social ambitions have generally been blunted, scaled back or even abandoned. Shifting to a social and ecological purpose does not necessarily mean the end of growth, but it places an organization on a different path, whereby its strategy and impact are honed and directed to do more than make money.

Dr Victoria Hurth, a colleague of mine at the Cambridge Institute for Sustainability Leadership, focuses on the connected themes of purpose, governance, marketing, leadership and culture. She observes that many organizations only innovate to achieve their own narrow financial goals, rather than to secure long-term well-being for all. 'This structures all your decision making to the point where nothing can get through the business case unless it can ultimately optimize your financial self-interest,' she says. Considering the scale and depth of the sustainability crisis, she wonders why organizations not working towards maximizing long-term well-being should have a right to exist at all. 'We can't solve these issues without enterprises being one hundred per cent focused on that, and that's what purpose-driven organizations are.'

Being solvent is obviously a core necessity for success in any organization, but it is not the only or main thing, she contends. 'You need enough financial stocks and flows to be healthy, but you also need the health of the non-financial capitals, the social and environmental systems and the stakeholders.' Each organization needs to identify its optimal strategic contribution, but all will need to work towards the ultimate purpose of long-term well-being for all, which is the goal of sustainability, she explains.

Hurth has helped create an official standard for purpose-driven organizations, helping companies to embed and enhance organizational purpose driven strategies, which was published in 2022. 'This provides the starting point to build consensus on what an organization whose decisions are aligned with long-term well-being for all looks like, and the next step is to create an international standard.' She describes how until recently the governance of organizations (which is basically what she is seeking to shift) 'grew out of stock exchange codes, which have been written to protect financial assets for shareholders'. Designing new codes and standards for organizations so that they fully embed social purpose amounts to a fundamental shift.

While this is a very big picture, one practical step organizations can take towards greater equality is to set a limit on how much the CEO is paid compared with their lowest-paid workers. When I took on the role of leading Friends of the Earth, the policy then was that the director would not be paid more than 3.5 times what the lowest-paid person earned. This was a clear and simple way of retaining some semblance of fairness. If private-sector organizations adopted a rule whereby the highest-paid person would, for example, be paid no more than ten times the lowest wage, there would be a clear incentive to lift the wages of those paid the least so that the most senior staff could earn more. This would help level income distribution across society, which is far from the picture at the moment, considering how the Trade Union Congress found that in 2021, in the UK, the median pay of a FTSE 100 CEO was (at £3.41 million) 109 times more than the median earnings of a full-time worker (£31,285).[9]

Companies shifting to the pursuit of long-term well-being for all, rather than the pursuit of profit for themselves, which is the case now, will be driven not only by enlightened leadership but also by investors and banks. While pollution, environmental destruction and human rights abuses were once accepted as part and parcel of how profit was generated, to do so now is a significant liability. Investors and banks need to recognize this and demand evidence of positive impacts from how they allocate money, and this is not least because of the rising risks to them of continuing with business as usual.

3. Financing the future

For centuries, finance has backed enterprise with investment, loans and insurance to assist the process of wealth creation. A spectacularly successful system has evolved, at least for those who have done well from it. But it is no longer viable, considering the rising pace of global heating, the scale of ecosystem degradation and the widening inequalities. Aligning finance with socially and ecologically positive outcomes is now an urgent priority.

I've worked with various organizations seeking to use finance to foster sustainable outcomes and I know that it's not easy to come up with strategies that match the scale of what is needed. It is, however, necessary. This is not only because of the urgency of ecological and social questions, it is also because of the rising financial risk coming from, among other things, new regulations that require reduced environmental damage, rules that demand greater disclosure of environmental impacts, public rejection of irresponsible brands, better businesses wiping out destructive ones (such as renewables displacing coal), and the effects of environmental degradation, including climate change-hit sectors such as supermarkets, as farming and food-supply chains are impacted by extreme weather.

It also makes perfect sense to direct finance towards sustainable enterprises when you consider the interests of the people whose money it is. Take the pensions being set aside by millions of people. A 30-year-old today, retiring in the 2050s or '60s, can expect to live in a different world by then, with outcomes later this century being determined right now, including the rate and scale of climate disruption. If pensions are all about providing for a more secure retirement, then profiting from fossil fuels, deforestation and pollution now cannot be part of rational investment planning.

I am an impact fund adviser with the Swiss bank Union Bancaire Privée (UBP). Impact funds are designed to create a positive social and environmental impact, not just a financial return (although that is sought too), through owning shares in companies that in this case are aligned with the UN's Sustainable Development Goals. This is far from straightforward and not a very exact science, but it does reveal that it is possible to secure a return for investors by putting money into companies that are part of the solution, not the problem (such as purpose-led companies driving sustainable agriculture, circular economies, renewable power or micro-finance in developing countries). Importantly, as part owners of the companies the fund invests in, we engage with their management to urge them to go further in their pursuit of environmental and social goals. In 2021, UBP launched a new investment vehicle that purchases shares in companies that are part of the solution for nature recovery. Our aim is to show what can be done, in the hope that others will copy the approach.

This kind of work will not be enough on its own, however. I asked sustainable finance leader Steve Waygood what wider change might embrace. Steve has spent 25 years wrestling with the question of how to align finance with the climate-change and nature emergencies, including working with Aviva Investors for many years. He spoke to me in a personal capacity about the case for change, and where it needs to occur.

Waygood is highly aware that the inequalities visible in environmental challenges, including the 'deep inequalities' between countries, are generating systemic risk on a huge scale. 'Some have got huge oil deposits, others gas, others coal, others rare earth metals, and you're starting to see that level of inequality play out diplomatically and if that fails the whole of our global economy fails if the physical risk runs away from us.'

The physical risk Waygood refers to is large-scale environmental change, which translates into risk for the financial system. The insurance industry in particular faces peril as losses due to global heating rapidly increase. 'The current structure of markets means that insurance will start to collapse. That means banking will unwind because it needs insurance to underwrite its collateral, and that means investments in banks and insurance companies will start to unwind too, and that's very definitely a huge financial stability problem, which could lead to systemic collapse, not just of the ecosystem or the climate system, but of the economic and financial system.'

It will be a market failure of epic proportions. As Waygood explains, this should concern us, because 'a market failure leads to a suboptimal outcome for society. That's the conventional definition of market failure. And if the suboptimal outcome is that the planet can no longer sustain economic activity or absorb the emissions that are being generated, I would contend that climate change and natural resource over-extraction are the biggest market failures that we've ever seen.' Having identified systemic risk, he advocates systemic change to meet it.

'The system itself needs to change, as economists would put it, to internalize the externalities, or make the polluter pay, and then valuation starts to be more accurate. It's a true cost, a fair cost, a justice-related cost. And then capital allocation starts to work properly.' He sees a vital role for the might of finance in advocating for change. 'We need to raise our game and our vision and speak truth to power at the top table, and by that I mean presidents, prime ministers, finance ministers and central bank governors, and make it clear that this stuff is their job too.'

Shifting our definition of growth, harnessing business for clearer social purposes, redirecting the financial system towards environmentally positive and socially equitable outcomes will need to be supported by changes to tax regimes to secure the additional public resources needed to support a just transition.

4. Raising the transition war chest
Richard Murphy is the founder of Tax Research UK and Professor of Accounting Practice at Sheffield University Management School. I first worked with him in 2007 on an initiative called the 'Green New Deal'. This foresaw the impending financial crash and proposed joined-up economic responses to it, including mobilizing government investment in renewable power and energy efficiency to create jobs, cut emissions and save people money, while providing an economic stimulus. It was very much a Keynesian idea, in large part inspired by US President Roosevelt's 1930s New Deal that set out to kick-start the American economy during the Great Depression. Murphy has spent a great deal of time in subsequent years researching how it would be possible to raise the money needed to finance such a programme, and he too draws parallels with the Second World War.

'We went into it with biplanes, and we came out with jet aircraft. In six years we moved forward by decades, because a lot of government money was spent on innovating,' he explained. With that parallel in mind, he believes an ecological transition could be used as an economic stimulus. 'The evidence is very strong that a significant slug of government money creates very large quantities of innovation and is likely to induce significant rises in employment. It will most likely provide a significant boost to business because it is a major recipient of the spending in question, and therefore we're not going to have an economic downturn. In fact, if anything, we'll have an economic upturn. We could actually create a virtuous upward circle in government spending and reducing debt by going green.'

He has researched and written extensively on how the billions needed for this transformation could be raised via different streams of taxation, including reforming how national insurance is charged for higher earners, increasing the corporation tax on larger firms and restricting pension tax relief to the basic rate of income tax. In common with many other advocates, he sees major potential from tax reform on wealth. He is sceptical, though, about such taxes being levied on possessions, such as homes, artworks, companies and cars, because of the difficulty of making fair valuations of these things, and the fact that, in any event, they would vary from year to year. However, he sees much potential in the income derived from wealth, such as rents, dividends and capital gains. 'There is no point trying to tax wealth, just tax the income from wealth and capital gains better and we could definitely collect the money that we need.' One way to do that, he says, is to equalize income tax and capital gains tax.

We also talked about how vast sums could be raised for social and environmental programmes without changing tax policies at all, but by gathering what's due, but which is presently not paid. In 2021–22 the UK tax authorities estimated that nearly £36 billion was going uncollected, amounting to nearly five per cent of all tax owed.[10] Murphy says that although it is a huge sum, the true number is likely higher. 'My personal estimate is that unpaid taxes are probably at least double the revenue estimate. So, eighty plus billion a year.' Murphy told me that collecting some of that money would be a sound investment. 'A person working for the revenue on investigations can recover ten to fifteen times their salary a year.'

He went on to explain to me how there are also opportunities linked with tax reform on savings. 'I've done academic research that shows how over 80 per cent of all savings in the UK are in tax-incentivized accounts.' He says that if the incentives were retargeted, huge sums could be redirected to transition programmes, new zero-carbon industries and regenerative farming, among other things. He estimates that if one quarter of new pension savings were incentivized by tax relief to go into companies delivering environmental solutions, an extra £35 billion per year would be invested into them. Billions more could come from creating what he calls a 'Green New Deal ISA' (Individual Savings Account). 'Seventy billion a year goes into ISAs, which is a pretty big number. Most people are astonished by that, there's over 700 billion in ISAs in total at the moment out of the total UK financial wealth of about 15 trillion.' Getting a proportion of that behind purpose-driven businesses delivering on socially positive, low-carbon and nature-recovery outcomes would be transformative.

Regarding fears that higher taxes on wealth would cause better-off people to leave the country, there is good reason to believe that the effect would be limited. One academic who has looked at this in depth and detail is Andy Summers, an associate professor of law at the London School of Economics. His research has looked at the difference between people who were born in the UK and grew up there and those who moved there from a different country.[11,12] 'The empirical evidence is that the vast majority of the population who were born and grew up in the UK, and have always lived in the UK, are extremely immobile; there's basically no evidence in a statistically significant sense that they would move in response to taxes,' he said, adding, 'The vast majority of taxpayers are in that group.'

He mentions the Laffer curve, which reveals the relationship between tax rates and how much revenue governments collect, showing that if tax rates go above a certain level, the amount available to government goes down because incentives to earn more drop. Summers told me that the evidence for this effect is often overstated. 'Even amongst migrants there's no evidence that if you tax them more so many of them would leave that it would cost you revenue. It would still be revenue raising, below a certain extreme point anyway.' With this and other insights derived from data, Summers concludes, 'The idea the wealthy will flee the country is really overblown.'

He speaks of London as especially attractive to wealthy people. He adds, though, that it is the cultural infrastructure that is most attractive and cites research involving in-depth discussions with rich individuals that reached some fascinating conclusions. 'Interviewees were very scathing, frankly, of people that moved for tax reasons, implying that they were unduly economically self-interested and not culturally sophisticated.' He saw that the interviews exposed 'quite a lot of stigma about tax havens being boring and culturally barren.'

On the strength of this and other research, Summers believes that 'one of the advantages of being rich is that you get to choose where to live, rather than where is the cheapest place. I think a lot of that's reflected in analysis that we've done suggesting that there isn't a huge share of people that are really trying to eke out the kind of highest post-tax income that they can, there are other things that they're taking into account.'

Other research that I've touched upon is relevant in this respect. More equal societies are better for everyone, including the rich; if the UK did wish to attract more high-earners and wealthy people, it might paradoxically be better for it to focus on becoming a fairer society and to invest in cultural institutions and a beautiful, clean environment.

The society that I experienced as a child during the 1960s and '70s was far more equal than it is today, in part because of the different approach to taxation. Although not popular among some of the super-rich, with the Beatles' song 'Taxman' one manifestation of the displeasure, high taxes on great wealth didn't cause an economic disaster, innovation didn't falter and investment wasn't driven away from the UK. Higher taxes on wealth enabled the National Health Service to grow, allowed children from working families to go to university, and provided the money for millions of social houses to be built. In the modern context, all of that still needs to be funded, while money must also be raised to address new systemic challenges now upon us. And while enjoying success from their chart-topping musical complaint, the Beatles didn't move abroad.

5. Switch subsidies

Alongside public services, some of the money raised by governments from taxation supports different industries. At present, however, the way government money shapes the economics of vital sectors, such as food and energy, can exacerbate environmental damage. For example, the IMF estimated that fossil fuel subsidies in 2022 totalled about seven trillion dollars, or 7.1 per cent of world GDP in that year.[13] This included explicit subsidies and the lack of proper charging for global warming and local air pollution impacts, the effects of both of which fall disproportionately on poorer people and create massive costs.

Unpriced externalities (that is, the costly damages caused by climate change and air pollution) can be charged for via carbon and other taxes, with the finance generated being directed towards pressing social priorities, such as fuel poverty. Additional resources could come from shifting the explicit subsidies, including tax breaks for fossil fuel companies. The case for intervention recently became even more compelling as oil and gas companies returned record profits in the wake of the disruption caused to energy markets by the Russian invasion of Ukraine.

As the shareholders of Shell, BP, Exxon and the rest reaped the massive financial windfall that came with high energy prices, many consumers in many countries couldn't afford to keep warm. This was not only down to high energy prices, it was also because when prices went through the roof, so, literally, did the heat that they'd paid for. An ambitious programme to improve home energy efficiency, particularly for people on low incomes living in rented properties, would save energy, cut emissions and improve health and well-being.

The Ukraine war hit the cost of food too, yet even with the rocketing prices in shops, many farmers are failing to make a living. At almost every turn growers have in the name of cheap food been encouraged and pressured to produce more at less cost, yet consumers and farmers are struggling. Large retail groups, meanwhile, have made handsome profits, and the owners of commodity trading companies, like Cargill, have accumulated fortunes. Flipping this equation so that farmers can prosper from providing consumers with good food produced in ways that restore the environment can in some part be achieved by redirecting public money.

A focus on soil health, reduced pesticide usage and the restoration of wildlife habitats in farmed landscapes is certainly achievable, and during my time as the chair of Natural England our organization has assisted the government in shifting public money towards the promotion of public goods, such as nature recovery and reduced pollution. More money directed to increasing the incentives for positive behaviour would help farmers go further and faster in producing food that doesn't cost the Earth, and could make sustainable food more affordable for consumers on low incomes. Using those same financial incentives to encourage land managers to open recovering countryside to more people for walking and outdoor enjoyment is an additional public good, and one which brings disproportionately bigger benefits to those suffering the most severe deprivation.

6. Adopt and implement transition plans for priority sectors

One of the reasons why it is so difficult to shift key industries towards more sustainable practices, even when money is invested to help them, is because jobs in greener industries are not being created in the places where the old ones are lost. In 2024, the announcement of the closure of blast furnaces at the Port Talbot steelworks in South Wales, in part because of modernization to lower-emissions technology, led to the loss of thousands of local jobs, with no obvious bridge to new employment for the skilled workers in the community who were suddenly left without livelihoods. This is one of the two last steelworks in the UK (the other one is at Scunthorpe), following the closure of multiple plants, at Corby and Teesside in England, Ebbw Vale in Wales and Ravenscraig in Scotland, among others. In the wake of the election of the new Labour government later that year, an announcement was made about seeing what could be done, but it would have been better to have had a transition plan in place before the job losses were announced.

The closure of coal mines across the UK in the 1980s also caused widespread hardships and was strongly resisted. While other jobs were eventually created in the energy sector, including in natural gas and, more recently, renewables, these were of little benefit for most of the communities where coal production was ended. It is no wonder, then, that organizations representing workers fought so hard against what was rightly seen as an existential threat to local economies. It was also resisted for reasons of culture and pride, as people who had worked the same coal seams for generations witnessed the fabric of their communities unravel. Over time, pride was increasingly displaced by poverty.

While some industries, like coal mining, are geographically constrained, others are more flexible. One up-and-coming sector is the modernization of buildings to meet today's energy-efficiency standards. A transition plan for the energy sector which prioritizes upgrading homes that presently waste energy would create many skilled jobs right across the country. It would also drive up demand for the manufacture of energy-saving products, such as insulation, energy-saving sensors and double glazing. The promotion of small-scale solar and heat pumps could drive additional employment and decarbonization, while enhancing national energy security. If those industries were built up nationally, positive synergies for job creation would accompany the transition to net zero. The new Labour government elected in July 2024 energetically embraced this narrative, hopefully laying firm foundations for its intended transition to zero carbon power by 2030.

Farming and fishing, two more sectors that have recently been subject to changes in policy to render practices more sustainable, have experienced profound disquiet. Involving battling against the elements, being out in all weathers, making high-risk judgements about where to fish or what to plant, not knowing until it's too late to change course whether they will make a return and thus some income, they can be lonely and stressful professions. They are made more stressful still when changes to policies and subsidies (lately motivated by environmental goals) cause fundamental shifts to the economic viability of food production.

As the world makes the transition to low-carbon societies while simultaneously helping nature recover, we'll need a better plan for how this transition will proceed, not only from a technological point of view, but also from the perspective of sustaining livelihoods and improving social well-being. Some of this is already happening, of course, but as affected communities quite understandably resist changes which they anticipate will harm them, without a more equitable plan in place, not only will many people be losers in the necessary transitions, but the process will go considerably more slowly than it otherwise might do, and, indeed, needs to do. A clearer plan that links social well-being, jobs and justice with the transition to a low-carbon and high-nature future will enable more rapid progress to be made. At one minute to midnight, this is an urgent priority.

As the speed of technological change gathers pace with the rise of artificial intelligence (AI), planning for fair transitions to renewable, sustainable, circular, regenerative and restorative practices is even more important. Anyone engaged in sustainability programmes will be struck by the prominence of AI applications among the solutions currently being proposed, and by how many of these might in turn lead to jobs being lost. New jobs will be created, but the transition will be smoother if it includes support for those who will be affected. And while we ponder the role of AI, we must, and as is the case with electric vehicles, think about the ecological impact of what is coming. AI, for example, uses vast quantities of power. Will that push up prices and make it even harder to escape from fossil fuels? And will control of that technology further concentrate wealth, or can it be harnessed in pursuit of more equitable outcomes?

Whatever the answers might be to those questions, transition planning must embrace building the skills base needed to underpin the new sustainable economy, not only in information technology, but in ecological design, nature-based solutions, development planning, engineering, the creation of a circular economy, renewable power and heat, architecture, transport and sustainable fishing and farming, among other things.

7. Fairer incentives for sustainable consumption

One way to combat the obstacles to environmental action arising from affordability concerns, while incentivizing people to consume fewer resources, is to charge for power and water consumption on a sliding scale, whereby modest use per unit is cheap, while heavy use is charged at more per unit. So, for example, an elderly person living on a state pension would pay very little per unit of power and water, whereas wealthier consumers would pay more per unit with rising consumption – if they want, say, to fill swimming pools or power larger inefficient homes.

Power and water utilities could adopt such measures quite easily. And if they were instructed by regulators to focus on the provision of water and energy services, rather than on selling units, their business models could be additionally transformed. This kind of shift in charging structure would need to go hand in hand with support for home energy and water efficiency gains, so that poorer consumers who can't help wasting energy because of badly insulated homes are not inadvertently hit by economic signals they can't respond to. Less wealthy consumers using a lot of energy for medical or welfare reasons, such as dialysis, would need protection too.

Better choices can also be incentivized among consumers and businesses via levies on some aspects of consumption. These include carbon taxes charged to companies to encourage them to invest in cleaner energy and more efficient operations. Such a pollution levy could help to reduce emissions by shifting investments to cleaner alternatives while raising money that governments could invest in, among other things, an international climate-change loss and damage fund. This would be a step towards climate justice and would help enable some of the most vulnerable communities to adapt to a fast-warming world. One recent estimate suggests that the cost incurred through each tonne of carbon emitted is US$225, putting a figure on what such a tax could logically seek to raise in relation to each unit of pollution.[14]

One principle that can be adopted in ecological taxation policies is to keep the overall tax the government receives the same, but raise more money from pollution and waste, and less from income. The idea has been around for some time, but too little used. One criticism of ecological taxation is that it is a temporary policy because the intent is to price pollution and waste out of existence, which means the revenue raised will hopefully go down. During the transition, though, it is a fair way to promote cleaner production and consumption.

Sectors such as aviation that face particular technological challenges in reaching carbon neutrality could benefit from specific policies, whereby, for example, one flight per year per customer could be free of a carbon levy, but the more flights taken by an individual, the bigger the pollution charge added to successive journeys. Frequent flyers would pay more, thereby encouraging fewer flights, while protecting the affordability of an annual holiday for working people and their families. A tax on the fuel used in private jets, if levied at the same level as paid at the pumps by motorists per litre on diesel or petrol, would in the UK alone raise nearly £6 billion per year. Revenues derived from aviation could be invested in alternatives to frequent flying, such as better broadband connections or transport options with lower emissions than planes – for example, high-speed rail to displace some short-haul flights.

Some of those with great wealth will choose to pay carbon taxes and other environmental levies in order to continue with their status-driven displays of the expensive trappings of luxury. For reasons explored in the last chapter, this would slow the transition down, so effectively banning the most excessive unnecessary consumption would be one way to signal the need for change. Without that kind of signal, the powerful cultural tendency to idolize and look up to the ultra-wealthy and their extreme consumption is likely to block progress. Media coverage of the superstar singer Taylor Swift taking a private-jet trip across the world to hug her boyfriend at the Super Bowl is very likely to undermine calls for greener behaviour, limiting the perceived value of leaving the car at home and walking the children to school.

8. Uphold rights to a clean and healthy environment
Pollution and environmental destruction continue because the entities causing the damage wield more rights and power than the communities they blight. With disadvantaged groups unable to stand up to powerful external interests, be they logging companies invading Indigenous lands or oil refineries locating their plants next to where people live, environmental damage continues in part because of a serious power imbalance. If the people being polluted had more influence, the clean-up of industry and the restoration of nature would be prioritized. This has certainly been my experience, including in Teesside in the 1990s, where when communities on the receiving end of pollution had access to information, and the influence that came from that, they were more effective in advocating for the clean-up of industry.

People exercising their right to be heard can make a difference on global issues too. Lately there has been a surge of legal cases being brought against companies and countries to demand justice for the damage being done by climate change. One report looking at legal activity on climate found that between 2017 and 2022 the total number of cases more than doubled from 884 to 2,180.[15] Many of these were in developed countries, but more and more legal action is coming from communities in developing nations, with a growing proportion also from children (including girls as young as seven and nine years old in Pakistan and India). Recently (2024) in Switzerland, plaintiffs successfully made their case based on the disproportionate impact of emissions on older women. Although individually most legal actions make a modest impact on emissions, more affected groups exercising their right to be heard will be a vital part of the transition, rebalancing the amount of power held by the polluters and the polluted.

Indigenous societies are also increasingly using the law to exercise their right to achieve fairer outcomes. Examples can be found in Argentina, Australia, Canada, Ecuador, New Zealand and the USA. The impact of these kinds of challenges can be expected to grow as clearer environmental rights become established. Indigenous groups also use legal defence to protect their right to speak out against those impacting on their lands. SAVE Rivers, the group we met in Chapter 4, facing a legal action from the Malaysian timber giant Samling, eventually won in their struggle to retain their right to speak out when the logging company finally dropped its claim against them in early 2024.

Another example of how local communities can impact on global challenges comes from Peruvian farmer Luciano Lliuya.[16] Close to his home high in the Cordillera Blanca in the Peruvian Andes, the accelerating rate of glacier retreat threatens devastation, as growing meltwater lakes could send a wall of water down valleys, destroying people's farms and homes without warning. In 2015 he filed a claim for compensation from the German energy giant RWE proportionate to the climate-changing pollution it had produced. The case, which is currently on appeal, has the potential to set a global precedent, raising the possibility that affected communities might be on the verge of holding companies accountable for the damage they are doing.

On top of the individual actions and cases, broader progress has been made too, including that seen in the unanimous vote at the UN General Assembly in 2022 backing a resolution which affirmed that a clean, healthy and sustainable environment is a human right for all, and not just a privilege (temporarily) enjoyed by the better-off.[17] Having noted that, in some countries the rights of citizens to protect the environment via protest have been curtailed, including through new laws. The long prison sentences handed down to Just Stop Oil campaigners in the UK reveal how the state can limit the ability of people to challenge the status quo, even when existential threats to civilization are being highlighted.

9. Use the law to protect future generations
The first country in the world to pass laws to protect the interests of future generations was also one of the first to industrialize. Shortly after England ramped up its coal-based industry, Wales followed, building vast industrial capacity on the plentiful supply of high-quality coal and iron-ore deposits that lay beneath the South Wales valleys. Huge wealth was generated. In 1904 at the Coal Exchange in Cardiff, the first ever million-pound cheque was signed. There is still a lot of coal left beneath the Welsh hills, but the prospect of it ever being combusted is now limited, in part because of a 2015 law that requires the government of Wales to take account of the interests of future generations in the decisions it makes.

The Well-being of Future Generations (Wales) Act 2015 requires the government and 48 public bodies in Wales to think about the long-term impact of their decisions in ways that will protect future well-being, while tackling persistent systemic problems such as poverty, health inequalities and climate change.[18] The ground-breaking Act seeks to advance seven well-being goals linked to prosperity, resilience, equality, health, community, culture and global responsibility. It has not been without its challenges, but the framework has already helped change the decisions and strategies of official bodies in Wales.

Of course, and as we have seen, protecting the interests of future generations will mean looking after the needs of those living now, otherwise change will continue to be blocked. Driving down environmental impacts while protecting the interests of those who are least well-off requires integrated plans and policies that can work together.

10. Joined up policies
A large part of the challenge we face is a consequence of how we tend to approach different subjects, whether climate change, pensions or health, as if they have nothing to do with each other.

Two of the most siloed areas are the environment and the economy. Seeing things in isolation leads, I have found, to a tendency to seek trade-offs between different priorities. This in turn encourages an assessment and judgement as to which is more important – for example, nature or economic growth? History reveals how that tends to pan out. But by adopting a mindset that seeks synergies rather than trade-offs, we can achieve better outcomes.

There are many examples of the potential for greater synergy between apparently conflicting goals. The Green New Deal mentioned above is one such proposal, combining social well-being, job creation, carbon reduction and energy security. Another is the provision of green space near to where people live. In England, green spaces bring health, climate-change and environmental benefits worth an estimated £6.6 billion per year, but one third of people don't have access to such places within 15 minutes' walk of where they live.[19]

This link between health and environmental quality has been recognized in a pioneering piece of work being undertaken by Wessex Water, which is seeking to understand how the provision of better green space might improve public health to the point where levels of metformin (a drug used to treat type 2 diabetes) might be diminished in the wastewater that the company needs to purify.[20,21,22] It is expensive to remove this chemical before the treated sewage is returned to the environment, so stopping it going into wastewater in the first place makes financial sense. The company is seeking to understand how savings could be made in part through improving access to good-quality green spaces.

This is a good example of the kind of integrated thinking we need. In my role leading Natural England I have found this kind of approach to be an essential starting point for making progress. This is because most of the things that our official government organization covers or touches, including farming, the state of rivers, the decline of wildlife, depleted tree cover, public enjoyment of green spaces, ecosystem carbon and resilience to climate-change impacts, are in one way or another all connected to each other.

Trying to make progress on any one of these in isolation will at best be suboptimal, whereas taking a system-level view, through which the connections and synergies between these themes are understood and pursued together, can lead to better outcomes, and also sometimes at less cost.

CRISIS AND INACTION

It is a very human tendency to carry on with business as usual until a major crisis is upon us, rather than acting ahead of it, even when we can see it coming. This is the situation we find ourselves in now, and, returning to parallels with wartime, it is comparable to that of Britain before the Second World War, in the years before my mother and tens of thousands of other children had to be evacuated from Britain's cities. Winston Churchill summed it up rather well in a speech he made in Parliament in November 1936. 'Owing to past neglect, in the face of the plainest warnings, we have now entered upon a period of danger . . . The era of procrastination, of half-measures, of soothing and baffling expedients, of delays, is coming to its close. In its place we are entering a period of consequences . . . We cannot avoid this period; we are in it now.'

Those exact words could be applied today to the deepening ecological crisis. In the face of the plainest warnings, and since the 1970s, '80s and '90s and the more recent period during which the alarm has been progressively more loudly raised, the world has more or less continued on the same historical path.[23] Soothing and baffling expedients, half measures, procrastination and delay have brought us to the brink of major disruption. When the consequences that Winston Churchill spoke of became apparent and the penny finally dropped in the wake of the Nazi invasion of Poland and then of France and the Low Countries, a crash programme of wartime preparations was put into motion, with social equality baked into the plans that were laid.

Although, unlike the Second World War, we cannot pin the start of the planetary ecological crisis to a particular date, it is well under way, and we are truly in the period of consequences.

The year 2023 was the warmest globally on record, and according to climatologists was 'virtually certain' to have seen the warmest temperature for more than 100,000 years.[24] That year also saw a series of unprecedented events, from extreme ocean heating to record droughts, and from intense forest fires to massive polar ice melting.[25] Then in 2024 the record was broken again, this time taking the global average temperature compared with the pre-industrial period to over 1.5 degrees of warming, with that extreme also contributing to the period from 2015 to 2024 being the warmest decade ever recorded. In January 2025 the UK Met Office reported that carbon dioxide concentrations in the atmosphere had jumped up at an unprecedentedly fast rate, due to the impact of record high emissions from fossil fuels magnified by massive wildfires and weaker uptake of carbon by ecosystems such as in tropical forests.[26] With these trends ongoing, we should during the years ahead expect more serious consequences for food and water security, mounting economic damage, mass migration and conflict. The parallel destruction of ecosystems will come with a series of additional consequences, magnifying the damage caused by climate change, while at the same time rendering people and our economic systems more vulnerable, with deepening social inequality both a cause and a consequence of all that. Despite the increasingly urgent need for action, however, the inertia that sustains the status quo retains huge momentum.

The question is not only what needs to change, it is also how to do it.

12

Thrivalism

Across decades of worsening reports about the epic pace of ecological damage, people like me have repeatedly written down lists of policies to fix it, just like the one in that last chapter. Proposals for concrete policies are an important part of what is needed, but history confirms that this will not be enough. The question of why this is the case remains crucial.

Part of the answer comes down to the fact that many of these proposals react to the most immediate and obvious aspects of environmental degradation, rather than considering what lies behind them. The result is that remedies to environmental challenges are couched as environmental policies, when it is in fact deeper economic, political and cultural questions that present the real conundrums we must grapple with, and of course the social context in which all that takes place.

The more I've thought about this, the more I see the challenge as a series of interconnected layers, perhaps like a geological formation, where it is impossible to understand the topsoil without knowing the character of the bedrock. In this case, the uppermost layer, that most of us see day to day, is the environmental science: the data and stories we hear about the scale and pace of global heating, the depletion of resources and the degradation of nature. We document those trends and their causes ever more thoroughly, proclaiming the need to act on fossil fuels, pollution, unsustainable farming and excessive consumption of resources. Environmentalists come up with killer facts to back their calls for action to cut carbon and protect more land and sea for nature. While none of that is incorrect, it is demonstrably ineffectual in achieving the pace and scale of change that is needed.

This is in part because beneath the environmental science and policy, there is a layer of economic priority. As we saw in Chapter 9, this is characterized by an obsession with growth. This shapes the layer above, with the promotion of economic development time after time trumping environmental progress. This is not the bottom layer though, for the economic layer is in turn shaped by politics, and the choices made by those who hold power. Their priorities, those of the interests that influence them, and what they believe citizens will support and vote for determine their agenda for policies and laws, which are generally explicitly pro-growth. Voters' views are in turn shaped by another layer: that of our culture.

Culture is very powerful. It embodies shared values, shapes how we do things, encapsulates social norms and determines behavioural choices. Culture is pervasive and for the most part we don't even notice it. It's like the air we breathe or the water that a fish swims in: omnipresent and taken utterly for granted. In the West, popular culture is dominated by consumerist values, and yet while culture is fundamental, there is a deeper level still that shapes it: our collective philosophical outlook.

In western countries this philosophical outlook, or worldview, is largely human-centric and detached from nature. The concept of a healthy natural world has morphed from something our ancestors would have assumed was a clear and obvious necessity into an abstraction that is 'nice to have'. With this perspective dominating the foundation of our social fabric, it is little wonder that it is so hard to make efforts to restore nature stick in the layers above, including at the top, where science and its implications for policy are discussed.

Throughout this book I have attempted to explain why social inequalities present major environmental challenges. It is undeniable that less privileged people are hit hardest by pollution, that the better-off use up far more resources, that poverty blocks action on environmental issues and encourages governments to settle for environmentally damaging economic ideas, and that the differences between people drive unsustainable and wasteful consumption. Progress has been held back not for a lack of facts or good policy options, but because of blockages in those deeper layers of economic ideas, political priorities and cultural norms, and because of our nature-detached worldview.

Yet when it comes to ecological questions, attention remains stubbornly focused on that top and most visible layer, with environmental priorities trapped in a superficial space where they consistently fall foul of deeper and more powerful forces. Calls to look after nature and the climate are consistently stymied by measures that instead promote growth in GDP, which sustains the culture of excessive consumption. This ideology is protected from serious challenge by a collective worldview which receives news of pressing ecological priorities as abstractions that are largely met with indifference.

As we enter the last moments during which it might be possible to avoid the very serious consequences of rapid global heating, ecosystem degradation and depletion of resources, we must urgently ask different questions. The key question is not about the optimum balance between ecological goals and growth in GDP, it is about the best ways to achieve well-being for all while at the same time repairing our planet's stressed and damaged life-support systems. One starting point could be the specific policies set out in the last chapter, but it is also necessary to adopt wider strategies that will take us more broadly in a positive direction. We need a vision that can encapsulate what is needed, together with plans for gaining the influence and power to deliver it.

When it comes to vision, the good news is that by acting to reverse environmental damage, inequalities can be addressed at the same time. For example, more efficient homes are cheaper to run and create a buffer against future energy price rises like the one seen in 2022. Access to green space has a greater beneficial impact on those suffering from the effects of social deprivation, over time saving billions of pounds on public health costs. More equal societies are healthier for everyone, not just the less well-off, but also those who are richer and more privileged.

By reducing the overall footprint of consumption, injustices linked with environmental impacts can be reduced. Restoring nature can help us all cope with climate change, including the most vulnerable communities at the sharp end of extreme events, and it can also help us protect food and water security, thereby making future positive impacts on the cost of living. Getting people to reconnect with nature right across societies would also help to create a mandate for the actions needed to halt and reverse the decline of life on Earth.

Scaling up climate-change mitigation strategies in the transport and power sectors offers the prospect of millions of new jobs, cleaner air and healthier people. The new skills needed for the transition would not only create quality employment, they would also drive innovation, new business models and opportunities for sustainable value creation. Healthy soils producing healthy food in landscapes where nature and people thrive alongside farm businesses could help to restore broken connections between food and people. A fairer society that shares environmental responsibility would rebuild social trust and cohesion, improving everyone's quality of life. Suffice to say, it is not hard to paint a compelling positive picture, one that sits above the tired trade-offs between growth and environmental imperatives. But it might help if we could give this vision a name.

Politics loves labels. While it might be tempting to file solutions to environmental degradation and inequality under the label of socialism or communism, it would be a mistake. This is not least because both ideologies come with serious environmental drawbacks. Take the former Communist Bloc countries, which were even more environmentally destructive than the capitalist West, wrecking ecosystems and polluting on an even grander scale. And in the West, where environmental questions have been more prominent, some left-wing thinkers have dismissed environmentalism as a middle-class preoccupation that is anti-poor in its intent, pushing up prices, destroying jobs and placing unnecessary burdens on the less well-off – the exact same narrative that has for decades so consistently blocked progress.

Capitalism too has revealed serious limitations, seen in the failure thus far to reconcile ecological limits within its core need for endless growth driven by the free will of capital. This has encouraged businesses to push back on regulatory measures to cut pollution and protect nature. Some ideological capitalists are climate-change deniers and resist environmental policies not only because they involve steps that will limit the ability of capital to offload environmental costs onto society at large, but also because they believe these policies limit personal freedoms.

It seems to me that the big ideas that have dominated thinking for many decades are no longer fit for purpose. We need a new way to describe the shift needed, one which lends itself to uniting questions of environmental sustainability and social progress. A new political frame is needed, going beyond the tired old 'left' and 'right' that so stubbornly constrains debates across most democracies. That frame might be encapsulated in the notion of *thrivalism*, a system that meets the needs of people thriving on a living planet, rather than consuming more and more without much consideration for the biosphere that sustains society and the economy, nor for how the benefits of progress are shared. The ten-point agenda I set out in the last chapter is one way of expressing the kinds of priorities it might embrace in seeking to promote well-being for all while reversing environmental damage.

Where could the momentum for embracing a new political philosophy come from? The answer is not one place, but many. Groups with the most to lose (and gain) will need to be involved, including younger generations who will, if decisive action is not taken soon, witness potentially disastrous environmental changes during their lifetimes. Those campaigning for racial equality will need to be engaged, as will the Indigenous peoples who retain the pre-industrial worldviews now needed more than ever in shaping modern perspectives. Women's movements need to participate too.

The power of their voices became clear to me when the UK's Women's Institute joined in the campaign for the 2008 Climate Change Act. But it is not only the disadvantaged and the directly affected who must speak up. Throughout my career in environmental and conservation work, people with wealth, connections and political power have provided critical support too.

SYSTEMS THINKING

Julia Davies is one of the UK's most active environmental philanthropists, using her wealth to make progress on nature recovery. 'As a teenager I got it. I was worried about what we were doing to the world; the way we were living, and the way society was going was completely unsustainable,' she told me. Growing up in South Wales, she had a modest start in life. Her father was a builder, but she went to university and trained as a commercial lawyer. She used that training to help set up a company called Osprey Europe, distributing outdoor backpacks.

Through Osprey Europe, Davies made a lot of money and set about using her wealth to make a difference on environmental issues. She also became involved with Tax Justice UK, a group campaigning for taxes on wealth as well as income, and then with the UK offshoot of a group founded in the USA called Patriotic Millionaires. For Davies, these two strands of activity – environmentalism on the one hand, and advocacy for fairer taxation on the other – are related.

'Extreme wealth is extremely corrupting of society, and I think that's really important. We absolutely have got to address this, and with Patriotic Millionaires UK we're trying to show a different way of being wealthy.' She described how, with a few people controlling extreme wealth, the inequality gap has just been getting worse. 'We had a period where things were getting better. We brought in the NHS, we brought in free education, we were supporting our young people. A young person born into a working-class family could go to university. We've been going back from that and that's obviously bad from a social perspective, but I do think that's interlinked to what we're doing to the world.'

Her remarks are a good example of the shift in thinking now needed, going beyond the tendency to focus on one subject at a time and looking at the wider picture. This is known as systems thinking. Fortunately, there are more and more people who, like Davies, see the need for joining the dots between, among other things, equality, ecology and the economy. It was this way of thinking that inspired her to become a campaigner for higher taxes for the very wealthy.

'Within Patriotic Millionaires we think the wealthiest in society should pay higher taxes. We have quoted back to us figures on the percentage of all taxes paid by the wealthiest [in] society as if that's supposed to shut us up. Well, sorry, no.' She explained how, for her, great wealth poses moral questions that need addressing via taxation rather than philanthropy. 'Jeff Bezos says he's going to give away most of his money in his lifetime. Well, "most" could mean just over 50 per cent, and "in his lifetime" could be that he's going to wait until he is 70 to do it.'

She went on to tell me how she thought Bezos, the founder and owner of Amazon, had 'structured his entire business operation on making sure that he's not really going to be paying much tax'. Davies said she believed 'his whole business model is based on putting other businesses out of business, businesses that do pay tax, like the little shops, and guys on a high street. He's massively taken away from the public purse with his entire business model. Then he wants us to think he's a good guy, because he's saying at some point during his lifetime, he's going to give away more than half of his mega fortune that's been made on the back of not properly paying tax, on the back of encouraging people to buy things they really don't need.'

Davies thinks this kind of attitude is supported by the malign influence of much of our media, which promotes unnecessary consumption and a very negative view of taxation. She has reached a similar conclusion to many of the academic researchers I spoke to. 'If we don't address this extreme level of inequality, then we are creating this society which is just breaking down. It breaks down communities, it breaks down trust, and if people don't trust each other, then they're less likely to look after each other and then you just have this sort of increasing spiral of people living hand to mouth, struggling.'

Davies's views on taxation are, contrary to what most newspapers might claim, quite widely supported. For example, in a 2022 survey by the UK's National Centre for Social Research, more than half (52 per cent) of the respondents backed increasing taxes, to be spent on health, education and other social programmes, with just under half (49 per cent) backing redistribution from the better-off to those with less.[1] It is not only in the UK where the mood among the very well-off is shifting to align with wider public views.

Morris Pearl of Patriotic Millionaires (the original US group) represents a network of some two hundred American members who believe inequality is a major threat to social stability. He rejects the idea that reducing taxes on rich people would help everyone. 'It's not true,' he said. 'Reducing taxes on rich people makes them richer. Period.' While he doesn't believe in 'trickle-down' economics, he does believe, by contrast, in 'trickle up'. 'If you give everyone else some money, that money quickly trickles up,' he said. 'People make their mortgage and rent payments and pay their phone bills and do all those things. Those of us who have the good fortune of having investments in the real estate companies that are their landlords, and in Verizon, that's the phone company, we become wealthier that way. We who are wealthy are dependent on the rest of the population having enough money to pay all these bills and be part of society. Sure, if somebody reduces my taxes today, I might have more money. But that's not going to help me in the long run.'

Pearl reckons many politicians favour the trickle-down narrative because of who they speak to. Having interacted with many US political candidates and funded some of them, he said, 'They spend so much time talking to rich people that they become very familiar with what the rich people have to say. They've not heard from a lot of people who work for a living. They've heard from investors and businesspeople and so that's the perspective they have, and after a while, they just believe it.'

Members of Patriotic Millionaires are convinced that change is needed to benefit not only the less well-off but also those who are more fortunate. 'This gross inequality is not good for rich people,' Pearl told me, explaining the priorities of the network. 'We talk about three issues. Taxation, that's where we started, and it's still one of our major issues. Sometimes we talk about raising the minimum wage, and the third thing we talk about is legislation to reduce the role of rich people in politics.'

It is not only individuals with lots of money who are influential in politics, of course, it is also financially powerful organizations. Finance expert Steve Waygood placed great emphasis on the potential power of banks, pension funds, insurance companies and others to shift political choices in a positive direction, not least because it is these organizations, and the financial markets they operate within, that enable governments to function.

'Companies with purpose and investors with purpose should be collectively engaging with the governments whose job it is to correct markets,' he told me. 'The power that sits within global banking, insurance, and investment, which lends over 100 trillion dollars to the world's governments, is huge. If the chief executives of these institutions recognize that it's in their interests, the interests of their shareholders, for market failures to be corrected, then I believe that power can be harnessed to actually encourage government to correct these failures. We need to get that to happen. We need the citizens who bank, insure and invest in these different institutions to encourage those businesses to use their voice.'

Asking millionaires and global capital to make the case for a more sustainable and equal world might seem like an unlikely route to positive change, but it's entirely logical. Continuing with business as usual brings the prospect of system collapse. We know they know this because some of them are making arrangements to ride out the disaster that they evidently believe is coming. One such is the billionaire founder of Facebook, Mark Zuckerberg, who is spending millions of dollars constructing an underground bunker in Hawaii. Complete with its own energy and food supply and a blast-proof door, it will enable one of the richest men on

Earth to ride out societal breakdown (temporarily at least). That's certainly one way to respond to what plausibly lies ahead. Another is to do as Waygood suggests and adopt strategies for durable and sustainable system change.

It would help if people in the top positions in politics, who would ideally be hearing from the rich and powerful about the need for change, themselves came from a wider range of backgrounds than is presently the case.

POLITICS AND CULTURE

Most of those who've reached high office in British politics have tended to be from better-off backgrounds, and thus have proven more inclined to reinforce the status quo. For example, in 2023 the Sutton Trust found that nearly two thirds of the British government cabinet (63 per cent) were privately educated, with more than half of them (52 per cent) attending the universities of Oxford or Cambridge.[2]

Of Britain's 57 prime ministers (as of 2023), more than a third (20) attended just one public school: Eton. Only 11 prime ministers (about 20 per cent) went to the kind of non-fee-paying schools that 93 per cent of voters attended.[3] There have been ups and downs in various countries in reducing some of the inequalities in politics linked with race and gender (the UK is one of them), but in many cases economic inequalities have grown progressively worse, and at the same time money has played an ever more important role in who has the best prospects for being elected. The result is that a relatively narrow cross section of society has tended to hold a disproportionate amount of political power and, for the most part, to carry the mindset and priorities of the privileged.

When it comes to gender diversity, there is evidence from developed countries that women take risks to the environment more seriously than men and are more likely to support strong climate-change policies. Women-led companies show stronger environmental performance, and women-led banks lend less to polluting industries and are more likely to invest in renewables.[4,5] This tendency is visible in politics as well. Countries with more female members of parliament are more likely to ratify environmental agreements, regardless of political leaning.[6] Anecdotally, it is interesting that it was two female British prime ministers who made what are among the most distinguished contributions to reversing climate change. One was Margaret Thatcher, who, in 1989, was the first prime minister to make a major speech on the need to act on climate change. The other was Theresa May, who, in 2019, introduced the world-leading legal target to drive down British emissions to net zero by 2050.

There is also evidence from several countries that stronger participation from women in politics is linked with lower levels of corruption.[7] This is important because corruption is one reason why there are sometimes insufficient public resources available to protect and restore the environment. Corrupt practices also destroy trust in official agencies, which makes it all the more challenging for politicians to confront ecological challenges.

And while there has been significant progress in British politics in terms of women and non-white people being elected, many of them have come from privileged backgrounds and championed the kinds of ideas that have led to the predicament in which we now find ourselves.[8,9]

In 2024 there was, however, a dramatic change. The landslide Labour victory in the general election that year recast the composition of the British Parliament, with the election of the largest number of female MPs ever, and a cabinet with the lowest level of privately educated people ever known.[10] It was also the most racially diverse parliament, with, for the first time ever, the number of non-white MPs reflecting the overall proportion of ethnic-minority people in the UK (about 14 per cent).[11]

Time will tell as to whether these shifts mark a permanent change of direction. In the meantime, the current parliamentary composition perhaps offers a unique opportunity for the consideration of deeper political reforms that might make it less likely that future parliaments and governments will revert to historical patterns.

When I stood as a candidate in a British general election, a few things struck me as potentially helpful in leading toward better outcomes. One was to favour the inclusion of more diverse voices via a more proportional electoral system, so that the overall number of votes cast is more closely reflected in the number of representatives from different parties that end up in parliament. At present, the UK's Westminster Parliament is elected via a 'first past the post' system that enables majority governments, nearly always comprised of one party, to be formed with less than a majority of votes. This system is quite rare globally, with far more countries adopting proportional approaches. The upper house of the British Parliament, the House of Lords, is comprised of members who are not voted into office at all, including a number who inherited the right to be there.

Another step that might help is making attendance at polling stations compulsory, so that everyone is required to take ownership of and responsibility for the democratic process. This would include putting a tick box on ballot papers for voters who don't wish to support any of the candidates who are standing. They would still need to actively state that though, rather than opting out of the process completely or spoiling their ballot paper (although they could still do the latter). Compulsory voting could help tackle trends evident in British elections whereby the most disadvantaged citizens are increasingly less likely to vote.[12] Young people are also significantly less likely than older voters to turn out, even though they will inherit the longer-term consequences of policies adopted (or not adopted). If these groups don't vote, their interests are less likely to be taken seriously by the state, which ends up perpetuating inequality. There is also a racial component to voter turnout, with white people more likely to vote than citizens who are Black or of Asian ancestry. It is noteworthy that the landslide election leading to the huge Labour majority in 2024 was achieved in the context of low voter turnout of 52 per cent.[13]

One example of how a more proportional system can help shift societies onto more sustainable trajectories was provided by the election of a group of loosely affiliated 'teal' independent candidates in Australia's 2022 federal elections. They stood on a centre-right 'blue' agenda combined with 'green' views on the environment – hence the name. The selection of the teals, all women, elevated climate-change action in politics, helping to transform the country's commitment on emissions reductions. This was in large part down to Australia's 'alternative vote' electoral system, which requires voters to rank candidates in order of preference and which enables candidates to be elected who would have little prospect in the kind of system used by the UK, the USA and many other countries, where just two parties tend to have most representatives. Australia also requires voters to turn up on election day, thereby guaranteeing a full turnout and in the process ensuring that whoever takes office has a clear mandate.

Another step would involve implementing tighter rules on political funding. Limiting the influence of the super-rich is evidently necessary, considering how it is the size of campaign budgets that helps to determine outcomes. Wealthy backers of parties are clearly more likely to seek policies that promote their interests, which in turn is less likely to lead to sustainable outcomes, given the huge role played by social inequalities in perpetuating ecological decline. Even so, changes to political systems are more likely to occur (disregarding revolutions) if they have the backing of wealthy and influential people. I was encouraged in this respect to learn of the work of Patriotic Millionaires in arguing for the urgent reform of political funding.

However, elections are quite rare events, normally taking place every four or five years, and for a successful transition to a more equitable system to work, it will be necessary to engage and involve people through more sophisticated decision-making processes during the years between elections. This will require a shift to more active citizenship, going beyond the current situation whereby most of us are regarded as consumers who occasionally vote. Giving people more influence via processes such as citizens' assemblies can be efficient and effective in enabling them to shape decisions, rather than leaving it all to the market or the power wielded by a tiny minority of political representatives. Seeking to harness the views of citizens brings us to the vital matter of the cultural context that shapes the expectations we have of political leaders. One source of influence in this respect is the media.

It is hard to nail down how the media might nudge cultural norms in sustainable directions, but now and again there are glimpses of how positive progress might be made. Julia Davies mentioned one such counterpoint to the cultural domination of consumerism. 'An interesting example of where the media is waking up to it and doing something different is what happened on *Love Island*,' she said, where the programme 'went from promoting fast fashion to promoting second-hand items from eBay.' It's a small example, but it does underline the different message that can be sent by a programme that shapes the cultural perspectives of millions of people.

Newspapers are another powerful way of shaping cultural norms. They often frame the popular discourse through the angles they take on the stories they choose to run and the opinion pieces they publish. Many of the main titles belong to a handful of billionaire press barons, who in turn have enormous influence over politicians and the policies elected leaders feel able to adopt. If the *Sun*, the *Daily Mail*, the *Daily Express* and other major papers don't like what political leaders have to say, then they are less likely to say it. At least, however, these mainstream titles are subject to some level of regulatory oversight in the form, in the UK, of the Independent Press Standards Organisation (IPSO). The rising power of social media is far less constrained.

That too is mostly owned by multi-billionaires, including the bunker-bound Mark Zuckerberg, who, with vast quantities of personal data harvested from countless online interactions, helps shape the context for debate on platforms that have been a growing influence on elections across the world. Via unedited feeds – where fiction masquerades as fact, anger is deliberately provoked, nuance is subsumed by outrage, conspiracy trumps reality and grievance is harnessed for political ends – those who enter these virtual echo chambers are widely influenced, manipulated and duped in pursuit of what are often unspoken agendas.

Social media is most potent where discord and inequality already prevail, and, in that context, when this wild west profusion of information is largely ungoverned, divisive populist themes emerge to shape political culture. The masters of these platforms harness division via algorithms that seek to generate more interaction, in the process driving up the advertising revenues that are a big part of their ever more extreme wealth. As societies are torn apart, vast fortunes are accumulated.

The consequences of alignment between on-line anger, political power, billionaire wealth and the control of major social media was laid bare during the 2024 US Presidential election campaign, when Elon Musk, the world's richest person, deployed his X platform to back Donald Trump. He subsequently became one of 13 billionaires to be appointed to positions in the new administration, marking a fusion of political power and wealth unprecedented in US history.[14] Musk also used the power of X during the summer of 2024 to attack the new British Prime Minister, stoking ill-feeling at a very sensitive moment, as racial violence exploded on the streets of the UK, including across some of the poorest and most deprived areas.

All this is far from positive and invites the question of what to do about it. We might begin to moderate the harmful downsides of social media by changing their status from platforms to publishers, in the process placing on them comparable responsibilities to those which apply to the proprietors and editors of newspapers and broadcast companies, requiring that the content they share is factually accurate and that news is differentiated from opinion. This is important because mainstream media and social media shape the ideas and the cultural context that determine the fundamental directions and outcomes pursued by societies.

Despite these downsides, the status quo is often defended as upholding freedom, including freedom of speech. Few disagree with the notion of freedom, but it is important to unpack the concept, and to realize that freedom for everyone to do as they wish can impact on other freedoms. This includes the freedom to enjoy a healthy environment. Julia Davies touched on this. 'While some complain about red tape and rules and regulations about what can be poured into rivers, I would say that we want to have the freedom to enjoy rivers we can swim in. So, what is framed sometimes as restriction is in fact protecting our freedom to enjoy a healthy lifestyle. I want to protect people's freedom to be able to walk down the street and breathe healthy clean air. I think framing is incredibly important,' she said.

The media and social media do a lot of that framing. When it comes to freedom, the difference often comes down to a distinction between 'freedom to' and 'freedom from'. 'Freedom to' might include the liberty to say anything that will achieve political advantage, even if untrue, or the freedom to pollute and consume without restriction. That can all be seen as freedom, but what about the importance of freedom from misinformation, or freedom from pollution or the consequences of climate change and ecological degradation?

This kind of cultural framing, which in turn determines the parameters of political debate, also comes from the marketing and promotional activities of companies. Victoria Hurth shared her views with me on this. She reminded me that because of business-as-usual short-term profit maximization, marketing 'has become very end of pipe, to get stuff out the door and sell it rather than meet people's needs, as it was positioned in the early days'. The undoubted success of the methods used to do this, and of how marketing has shaped culture, leads Hurth to conclude that it needs now to be part of the solution, harnessing huge marketing budgets to assist with the cultural rewiring needed to achieve sustainable outcomes. This is in part, she said, about appealing differently to people's multiple identities.

The dominant identity frame harnessed by marketeers and advertisers is a very familiar one, she said. 'Being affluent has become symbolically connected with success and good looks. It's a hugely powerful identity and [it's] wielded by marketeers.' She explained how, by contrast, 'the environmentalist identity is essentially wielded as the opposite'. This has become a problem for environmental advocates. As Hurth put it, 'Not only do we anchor affluence to consuming a lot, but then we anchor environmentalism to poverty, and if we do that we've lost, and that's what we mainly do. Instead, we need to rewire what it means to be affluent away from consumption.' In other words, the challenge isn't so much about making environmentalism more like the current view of affluence, it's about changing what we mean by the affluent identity, redefining the mainstream cultural view of the good life. That is a big thing to bring about.

At the moment, 'going green' is often equated with the opposite of being fashionable and doing well, and thus it struggles to gain positive cultural salience. This is a major problem, but considering the brain power available to the communication and creative sectors, it should not be insurmountable. After all, if product and brand promotion can get us to buy endless unnecessary household clutter, it should not be beyond the capability of gifted influencers to find ways to make a positive sell on a sustainable future, especially when it is in the direct long-term interests of the organizations they work for.

In addition to the marketing profession, Hurth reflected on the power of individuals in influencing culture. 'Everybody has a role in shaping and constructing the world around us. Every time we buy or don't buy something we reinforce it, we show it, we wear it, we embody it. The more you have an identity, the more you will have power to construct the world around you.' I was struck by her point about seeing not buying as a powerful statement, a viewpoint that often struggles to gain traction as the cultural focus remains stubbornly fixed on buying different, rather than not buying at all.

Changes to political systems and the cultural norms that shape them might help bring a wider range of views into the institutions that govern societies while at the same time shifting our collective desires and aspirations, but if we are to make more progress, we will need to harness insight from an even more fundamental place.

WISDOM

In a world full of data, wisdom has become surprisingly scarce. It is not extinct though. It holds on in pockets, including among those whose worldviews predate the modern age of consumerism, endless economic growth and disconnection from the natural world. It is to be found among those Indigenous societies that still seek to live in harmony with nature.

According to the International Labour Organization and researchers in conservation, Indigenous peoples make up about six per cent of the world's population and yet manage about one quarter of the world's land, which intersects with about 40 per cent of all our terrestrial protected areas and intact landscapes.[15,16] These people really are on the front line, even though they are among the most impoverished on Earth, making up 19 per cent of the world's extreme poor.

Centuries of oppression, exclusion and persecution and the toll of disease decimated their populations, and yet in many countries Indigenous societies managed to survive, including in the former colonies of Great Britain and other imperial powers. In Australia, Canada, New Zealand, Brazil, Bolivia, Ecuador, Colombia, Peru and other countries there has during recent decades been a rise of the Indigenous profile in politics. This is very important because, by contrast with the European colonists who came to dominate the world from the 1400s onwards, the native peoples they encountered had an utterly different relationship with the world around them.

They were, and many still are, embedded in nature. In stark contrast with those who came in search of gold, gems, furs, timber, slaves, land, empire and trade, their cultures and societies were closely entwined with the natural world. Many revered and worshipped the ecosystems and animals that sustained them. Their myths, legends and beliefs were nature-centric, not derived from a remote Christian God who'd become increasingly detached from Creation. They saw their well-being and the continuity of their peoples as dependent on nature that was thriving and intact, not on the mastery of technologies or control over resources. The voice and the philosophy they can bring to twenty-first-century challenges are thus vital for resurrecting, nurturing and empowering.

During the many years that I've been involved with environmental questions I have been repeatedly struck by the stark divergence between western and Indigenous worldviews, and increasingly encouraged by the extent to which the latter have been able to complement the former. Halting the progressive destruction of the Amazon rainforests has been a cause célèbre for decades – and with good reason, for if deforestation there is not halted and reversed, it will not be possible to avoid dangerous global heating, or to stop the impending mass extinction of animals and plants. I've been involved with that struggle in several different ways.

I've lobbied governments to crack down on illegal forest clearance, organized consumer boycotts of rainforest timber – most notably mahogany extracted from Indigenous territories – advocated new national parks, sought to influence the funding priorities of international agencies and worked on plans to pay countries for the carbon they keep in these amazing ecosystems. Some of that has made a difference, but not as much as when Indigenous people have gained control of the land.

Maps of the Amazon rainforests showing the overlap between Indigenous territories and where the forest remains confirms the link between their pro-nature outlook and what happens to critical ecosystems.[17] There has also been a growing positive impact at the global level, with Indigenous voices lately becoming more impactful in the negotiation of the international accords which, if we are to get onto a sustainable track, will need to be implemented in every country on Earth.

A case in point was the negotiation of the Global Biodiversity Framework agreed at the Montreal summit in 2022. I was there as part of the UK government delegation and was very encouraged to see strong, clear and impactful Indigenous voices shaping the talks. Thirty years previously, in the run-up to the Earth Summit in Rio de Janeiro, I saw how some countries, including Brazil and Malaysia, resisted even having Indigenous people mentioned in the agreements under negotiation, fearful of political control shifting away from governments and instead to people who laid ancestral claim to much of these countries' land and resources. Previously sceptical countries have over time, however, softened their approach, and in the new international agreement that was signed by nearly all the countries on Earth (and the UK team helped ensure it was a strong one), the role of Indigenous knowledge in adopting the means to conserve and use nature sustainably is firmly recognized.

Two years later, at COP16 in Cali Colombia, I joined the UK government team who helped to negotiate a new agreement that will enable stronger and more meaningful participation of Indigenous groups and other local communities in the work of the convention. This is very significant, because by providing for a more formal route for their perspectives to be included, their more profound nature-centric worldviews can help to shape a different discussion.

Eli Enns, the Canadian Indigenous leader whom we met in Chapter 9 and who is president of the Canadian section of the International Union for Conservation of Nature, was at the Montreal negotiations. He told me that through the teaching of his elders he'd come to understand how there are 'elder societies or elder civilizations and then there are younger societies, juvenile civilizations'. He explained that the idea of juvenile civilizations was not derogatory, but a reflection of how some were new, being a few hundred years old or less, and others had evolved their worldview over thousands of years.

He went on to explain how younger civilizations lacked wisdom about nature – seen, for example, in how European settlers who arrived in Canada built towns in places prone to flooding and ploughed prairie grasslands, eventually turning the landscape to a dustbowl. He told me how, having broken the land, the newcomers doubled down with their industrial methods. 'They started using chemical fertilizers, pesticides and herbicides, and they started intensively cultivating cow, pig and chicken populations in close quarters, which led to a focusing of effluent and the need for antibiotics to control the health and maturation of these creatures.' Enns told me that the ecological disaster that has hit North America since colonization was foreseen by Indigenous ancestors. 'They had visions of great harm that would come not only to their own people, but to the Earth.'

They also predicted, though, that the great, great grandchildren of the newcomers would, when the world was on the brink of destruction, seek guidance from the great, great grandchildren of the Indigenous people with whom their ancestors had made first contact. Enns told me that this for him meant two things. 'Number one, no matter what happens we have to really keep alive our teachings on how to live the good life and how to live a life in balance with the rest of Creation. And number two, probably the harder one, is that when the newcomers' great, great grandchildren come to our great, great grandchildren, we must as much as possible have an unpolluted heart and have forgiveness, and then work together to get out of that problem.'

Another Indigenous leader in Montreal seeking to influence the negotiations was Tatiana Degai, from the Itelmen people who have inhabited the Kamchatka peninsula in eastern Russia for millennia. She spoke at the summit about the beliefs of her people, including their Creation story, which is the source of their name, the Children of the Raven.

According to their tradition, thousands of years ago a raven was flying above the ocean and saw a huge king salmon in the sea, so he caught it with his claws, and as he dragged it up, the salmon turned into the land. 'Then he created hills and rivers and infused his hot spirit into the mountains. Then the raven believed the land should have life so populated the rivers with salmon, smelt and other fish, and then whales and other animals. Then he created the humans and taught them how to live in balance with the land and its animals.' The story tells of how everything is connected and how there are no people without the animals, no humans without the land, the oceans or the mountains, the rivers or the salmon, and no humans without the raven. 'If you take a piece out of this system, the whole thing can collapse, and this is what we observe now,' she said.

She spoke of the intergenerational transfer of knowledge and how that is conveyed in their native languages which are 'the original voices of the Earth'. The collective cultural memory that enables such people to live in harmony with their non-human relatives is a golden thread that traces ideas back to prehistory. 'We do not govern nature and the concept of people above nature does not exist in our way of living. We have a relationship that is built on equal partnership.'

Contrast that worldview with the prevailing western perspective, which seeks more economic growth and more consumption and, in the face of the plainest evidence, continues to destroy the very natural systems which ecological science has revealed sustain civilization. Enabling more influence from Indigenous voices presents a huge opportunity, through full social inclusion, participation and political representation, as well as through control over ancestral lands.

There are signs of progress. For example, in Brazil in September 2023 a landmark judgement was handed down by the federal supreme court of Brazil granting to the Xokleng people of Santa Catarina control of land that they had been fighting to gain legal rights over for more than a century.[18] This was an important moment for the Xokleng, but also for many other Indigenous groups because the judgement has major implications for how they can now exercise their rights in making land claims. Xokleng leader Yoko Kopacá explained how she saw the struggle. 'We are not defenders of Nature, we are Nature defending itself.' Such is their philosophical connection with the ecosystem that sustains them that they consider themselves part of it. Most western people don't think like that, but if they did, even somewhat, it would be transformative.

After the Montreal biodiversity summit King Charles hosted a meeting at Buckingham Palace to encourage swift implementation of the accord that had been thrashed out. A long-standing advocate for greater attention to native peoples' wisdom, the King had invited several Indigenous leaders, among them Domingo Peas, the Achuar Indigenous leader whom we met in Chapter 4 and who is fighting to save the Amazon rainforests from oil and gas companies. When I spoke to Domingo, I heard a similar perspective to that of other Indigenous leaders. 'We Indigenous peoples, we are the forest. That's why most of the Amazon is still intact. The majority of the rainforest that still exists is in our Indigenous territories under our care. The rest of the forest that's not Indigenous land is heavily threatened and is being destroyed rapidly.'

His view of the forest is also very different to that of the leaders of the businesses seeking resources from the land. 'The biggest challenge facing all of the Amazon Basin, not just our area but the entire Amazon Basin, is the great big transnational corporations that are seeking commodities like timber, oil and mining, and this, in the name of this grand vision called economic development, is threatening our forests.' He emphasized the need for greater awareness. 'We must create consciousness for people to understand that we cannot lose the Amazon. We need the Amazon for the world.'

This is not to say that his people are isolationist. They do want to embrace benefits from modern technology. I saw an example of this when visiting Indigenous communities in the Peruvian Amazon in 2017. I went there with the rainforest organization Cool Earth and met Ashaninka people using camera traps to photograph rare animals like tapirs and jaguars, and uploading the images via satellite connection to participate in wildlife surveys. In so doing, they were using technology to generate data and raise awareness in ongoing work to protect the forests. The Ashaninka and other Indigenous groups are also using GPS to demarcate their land and to document and publicize cases of illegal incursion by loggers and others. I was struck by the power unleashed when Indigenous knowledge was combined with modern technologies.

If Peas and others are to succeed in saving their lands, then attaining stronger representation in national life will be vital. 'It's important that we Indigenous peoples have a voice and a presence at the political arena, incorporating our vision and needs and also regulating and legislating, creating enforceable laws.' Peas believes, as other Indigenous leaders do, that they will not be able to do this alone. 'All sectors of the society need to be creating a new system and new agreement around the future of protecting areas of high biodiversity like the Amazon, and we can't do this in silos.'

Domingo Peas is one among a great many Indigenous leaders who offer common cause and partnership in making the transition to a sustainable society. Eli Enns, who told me that elders alive today believe we are now at the time when it was predicted the great, great grandchildren of colonists and Indigenous people would work together, explained how, through 'social innovations we can work collectively together in ethical space to find mutually beneficial solutions to these complex problems'. As part of that process, he believes that 'science is needed, but to go hand in hand with Indigenous knowledge in order for there to be balance'. Tatiana from the Itelmen people offered partnership too, in sharing knowledge and working in common cause to protect and restore the Earth's damaged life-support systems.

Listening to these stories from Indigenous leaders, and hearing of their nature-centric philosophy and their wish to find common cause with societies that formerly colonized their territories, it's clear to me that those with influence in western society have opportunities to join with some of the poorest people on Earth to secure solutions to pressing ecological challenges.

Empowering these presently disadvantaged, often excluded and marginalized communities is a priority. Matthew Owen is the director at Cool Earth, where Indigenous empowerment is at the core of the organization's strategy for protecting the remaining rainforests. 'The whole world has a critical dependence on Indigenous peoples to keep rainforests alive and yet it is, absurdly, still largely treated as a niche issue,' he said. He explained how the Indigenous cause gets 'maybe half a percent of climate funding', even though their communities control almost half of intact forests, 'and if we're serious about climate and biodiversity we need to do everything we can to dial up their resilience and minimize their vulnerabilities'.

As chair of Cool Earth, I am excited by a new programme being led by Owen and our team that makes no-strings-attached cash transfers to Indigenous people in the Amazon, Papua New Guinea and the Congo Basin. The idea is that they will know best what to do with the money, how to sustain their communities and organize the legal work needed to gain control of their territories, so putting them in the driving seat is logically going to lead to the maximum positive impact. The money is transferred at village level, and also to households, with mothers the principal recipients. 'The less we do, the better,' says Owen. 'We need to empower people and get out of their way.'

Someone who really understands the pivotal importance of Indigenous societies has ancestors who enabled the expansion of the empire during centuries gone by. At the COP28 climate change summit in Dubai, King Charles said in his powerful speech, 'We need to remember too that the Indigenous worldview teaches us that we are all connected, not only as human beings, but with all living things and all that sustains life, [and] as part of this grand and sacred system, harmony with Nature must be maintained. The Earth does not belong to us, we belong to the Earth.'

His words were very much in the spirit of cooperation that I'd heard from Indigenous groups, and I hope that his calls for harmony based on Indigenous wisdom will be heeded by the governments, financial institutions and companies that gather at global summits, and that in the end possess the power to change course. Harnessing indigenous wisdom is also of course about public opinion in the countries where Indigenous societies remain, especially their non-indigenous people.

A reminder of how far there is still to go came in 2023 with Australia's referendum on the question of whether to create clearer rights for Aboriginal people. Voters were asked to approve changes to the Australian Constitution that would create a body to be called the 'Aboriginal and Torres Strait Islander Voice', that would have been empowered to make representations to Parliament and the government on matters affecting native peoples. The popular vote across all of Australia's states rejected this proposal, leaving the country's Indigenous societies with no formal rights in the political process, even though they have been on the land there for about 65,000 years longer than the British and other colonists who began to settle that vast continent less than two and a half centuries ago. Considering that history, however, the Aboriginal people are well equipped to take the long view and are unlikely to regard that setback as the end of the story.

ACTION AT ALL LEVELS

In this chapter I have rather breathlessly covered a lot of ground, brought many ideas together and described multiple facets of the challenge at hand, and in so doing, I have set out the great breadth of what I believe needs to be addressed. As I have done this, my thoughts have been drawn back to 1990 and the time I went to work at Friends of the Earth. My new boss then was the charismatic campaigns director Andrew Lees. He was very clear in his view that our work needed to foster what he called 'action at all levels'.

His point was that it would not be enough to focus on only one priority at a time, but that we would need an ambitious political strategy to raise public awareness, persuade business, influence culture, develop good policy ideas, mobilize the grass roots and build a global movement all at once, not one at a time. This was because he had concluded that none of these things would be sufficient on its own. The result was a challenging and at times chaotic working environment, but it made sense and, more importantly, it made an impact above and beyond many other approaches I've seen down the years, where carefully laid, specific plans have created clarity, but often haven't made much difference.

Applying that kind of thinking to the themes explored in this book leads me to suggest that there is no one thing that will create the system-level shift needed. Instead, networks of organizations and leaders will need to collaborate to build a multilayered programme of change, working simultaneously at different levels. Collaboration between leading businesses, financial institutions, influential individuals, academics, policymakers, campaigning organizations, Indigenous groups, activists, political figures, think tanks, philosophers, writers, musicians, artists, economists, scientists, marketeers and commentators could place national and global conversations onto a new track if they converged on broad ideas that recognize how environmental and social challenges are deeply connected with one another, not separate, and that integrated approaches are needed to meet them.

This sounds like a very big ask, and it is. But it is possible. We know it's possible because it's happened before. If we look at the origins of the current system, during the 1930s neoclassical economists found common cause with politicians, think tanks, businesses, investors, newspaper columnists and academics to create and propagate the dominant neoliberal ideology which today defines economics, determines policy and shapes culture across so many countries. It didn't appear by accident, but instead was the result of a deliberate programme that prevailed because it was based on action at all levels. What is needed now is for a new idea to gain hold, based on new measures of progress, the fair allocation of resources, respect for ecological boundaries and opportunities for all, and with a focus on quality of life for everyone. Thrivalism is a one-word way to describe it.

Crystallizing a compelling alternative to the ecologically doomed programme to which we are presently committed is an urgent priority. Replacing the incumbent system with a new one will require the kind of broad-based collaboration that enabled the world we live in now to become established in the first place. It will need policy ideas, political support and social narratives that strengthen bonds across society to create the cultural norms that will enable people to see themselves as part of a common endeavour. It is worth noting that it took several decades for the movement that eventually succeeded in embedding neoliberal ideas to build support for its decisive breakthroughs. It's now more than 30 years since the Earth summit, and it might be that the opportunities for a new idea are approaching a similar point.

When I spoke to Kate Pickett, she shared with me a positive reflection as to what might happen should we succeed in tackling the deep-seated differences which lie at the heart of our ecological crisis. 'The message from our research is that your quality of life will be better if we reduce inequality, so we can tackle the climate change better if we reduce inequality, but that won't make life worse for you, [it will] actually make life better.' With that kind of thought at the core of a new idea it should be possible to build support, if those who see the need for system-level change can cooperate like the people who created and maintain the current system did.

With the alarm raised, targets set and a transition under way, one place to observe progress is back where we started, in that crucible of the Industrial Revolution on the river Tees, where both heavy industry and empire have receded into history. The Redcar steelworks and its towering brick chimneys have been demolished, the forest of factory pipes that once sent forth a multitude of chemicals largely gone. The air is cleaner and the blackened and polluted landscape that was bereft of wildlife is once more turning green.

Teesside's era of coal, chemicals and steel has ended, and now twenty-first-century technologies are on the rise. Wind turbine facilities service expanding offshore renewable power, and there are plans for a hydrogen hub as well as for carbon capture and storage infrastructure. The plan is to create a cluster comparable to that which emerged in the 1820s, but this time it will be based on clean low-carbon technologies. Nature recovery projects are also under way, with birds, mammals and fish returning to the lower Tees Valley following a long absence.

Once host to the first Industrial Revolution, Teesside now strives to participate in the Sustainability Revolution. But although progress is being made to cut Britain's carbon emissions, the social inequalities that remain are a stark reminder that we need more than clean technologies for truly sustainable outcomes. In 2019, Middlesbrough was ranked as England's fifth most deprived place (out of 317). Nearly two in five children there grow up in poverty, wages are ten per cent lower than the national average and poor health leads to lower life expectancy than in the rest of the UK.

Shahda Khan is the director of a Middlesbrough community programme called Borderlands. She was born in Kashmir and came to England when she was just a year old, moving to Teesside at 16. She and her team facilitate various discussions with local people, including some on nature and accessing green spaces. Her vision is one in which local people drive the change for themselves, not having it dictated to them or handed over as a *fait accompli*.

Shahda explained to me how one backdrop to Borderlands' work is serious deprivation and inequality, including 'huge health inequalities' and 'some of the highest rates of male suicide in the country'. At one point, Middlesbrough also had the highest number of refugees and asylum seekers in the entire country. 'The way that the Home Office works it out is to say that only one in every 200 members of settled communities should be a refugee or asylum seeker. In Middlesbrough at one point in two of our wards it was one in every 17. That was based on the cheap housing that was available locally, which again is the inequality.' She connects this to a wider global picture in which people are living in degraded environments not of their making. 'This country is one of the most nature-depleted countries in the world and we are having conversations with communities here saying it's not necessarily you who's causing all this damage, but you're going to feel the impact of it.' Another backdrop she sees is a theme in public debate which somehow blames the poor for their plight. 'There's that whole demonization of the poor . . . going back to the Victorian era of the deserving and undeserving poor.'

Working locally at grassroots level, Shahda agrees that there is a pressing need for a long-term vision. 'We don't need a five-year strategy or a ten-year strategy, we need a one-hundred-year strategy,' she says. 'That is how we should be thinking, that's how things change. There's too much short-termism when we need intergenerational change. What are we leaving for our children and grandchildren?'

And that seems like a good place to conclude, with the thoughts of a British Pakistani woman from Kashmir, a region subjected to violent division as the former British Empire broke up. Living in one of England's most deprived cities and working to connect people with natural areas in a place long blighted by industrialization, Shahda speaks about the links between ecological challenges and inequality, and makes the case for intergenerational responsibility. As she suggests, we don't require a tweak to this policy or that one, we need a long-term plan that confronts environmental and social questions together.

The Industrial Revolution and the post-war growth boom were both revolutions led by people of vision, and we need similarly transformative leadership now, to ensure the health, happiness and security of the ten billion or more people expected to be living on Earth at the end of this century. Carrying on as we are will chart a continuing course towards ecological disaster and mass extinction. But if we take decisive action right away, embedding system shifts that drive ecological recovery in tandem with greater fairness, it might still be possible to navigate the challenges that face us on our continuing journey into the Anthropocene. This is not a conclusion derived from ideology, it is a necessity born of the practical fact that we have nowhere else to go. There is just Earth.

Acknowledgements

Just Earth was made possible through the assistance of a great many friends, colleagues and experts from across a range of fields. I am especially grateful to my friend of many years Laura Fox, who was my research assistant for this project, helping with arranging and transcribing interviews, commenting on drafts and sourcing and organising the referenced material.

The book was written following a discussion with Nigel Newton, the Chief Executive and Founder of Bloomsbury Publishing. I am very grateful to him for seeing the sense of the idea of *Just Earth* and to his team for working with my agent Caroline Michel and her colleagues at Peters Fraser and Dunlop to make arrangements for it to be published. My editor at Bloomsbury, Tomasz Hoskins, helped with discussions about the structure and scope of the book and it was a pleasure to work with his colleague Octavia Stocker, who did an excellent job editing the manuscript. I am also very grateful to Fahmida Ahmed who oversaw the production of the book, the careful work of Mandy Woods, who undertook the copy edit and to Lora Findlay who produced the cover design. Colin Hynson completed the index and Nick Avery prepared the graphic figures that appear through the book. I am also grateful to Rachel Nicholson and her efforts to generate publicity for the book. Dr Peter Brotherton, my colleague at Natural England where he is Director of Science, came up with the rather brilliant title of *Just Earth*.

Several friends and colleagues were kind enough to read drafts and to provide feedback. Nick Rowley, former climate change advisor to Tony Blair while he was UK Prime Minister, gave many valuable detailed comments, many of which are reflected in the book. Katie Hill of *My Green Pod* read through and flagged a number of important points subsequently captured in the final book, as did Rosamunde Almond, who is among other things editor of WWF's *Living Planet Report*. Mike Childs, a long-serving Friends of the Earth campaigner and with whom I worked for many years, was good enough to take the time to read a draft and provide feedback. Christina Coleman helped with some early thinking on scope and direction and Rory Sullivan at the Grantham Research Institute on Climate Change and the Environment also provided feedback on an early draft. Faye Sommerville contributed additional research material and I would like to acknowledge the encouragement received during the early stages of this project from my colleague Alice Spencer at the Cambridge Institute for Sustainability Leadership.

I am immensely grateful to the many experts who spared time to be interviewed. Professor Johan Rockstrom at the University of Stockholm and Professor Sir Bob Watson shared their views about climate change and nature emergencies. Professor Susan Smith and Professor Simon Szreter at the University of Cambridge shared insights from their long academic careers studying social inequalities, as did University of York Professors Kate Pickett and Richard Wilkinson.

I was pleased to speak with Tim Lang, Professor Emeritus at City, University of London, who provided me with important insights on the interactions between food policy and inequality. Professor Sean Nixon at the Department of Sociology at the University of Essex told me about his work researching the origins of advertising and Dr Victoria Hurth, my friend and colleague at the Cambridge Institute for Sustainability Leadership, set out thinking for me about her research into purpose driven organisations and the role of marketing in shaping sustainable outcomes.

Professor Richard Murphy of Sheffield University Management School and founder of Tax Research UK, and Dr Andy Summers, Associate Professor of Law at the London School of Economics, shared their research on progressive taxation. The book was further enriched through conversation with Professor Tim Jackson, who leads the Centre for the Understanding of Sustainable Prosperity (CUSP) at the University of Surrey, and sustainable finance leader Steve Waygood. I am also grateful to Victoria Leggett, Head of Impact Investing at Union Bancaire Privee, for thoughts she shared with me about investment strategies to promote sustainable outcomes.

My former colleagues at Friends of the Earth, Craig Bennett, Elaine Gilligan, Mary Taylor, Ron Bailey and Mike Childs, provided interviews, enriched my recollections and pointed me toward source material that strengthened the book. I was pleased to speak with Danny Sriskandarajah, former Director of Oxfam UK, about his work to investigate the scale and implications of rising inequalities and to Shahda Khan, the director of Middlesbrough community programme Borderlands, on her experiences on the interactions between environmental questions and social inequalities. I was also very pleased to speak with Rhiane Fatinikun, the founder of Black Girls Hike, to discuss questions of race and how that shapes access to natural areas.

I was delighted to speak with Aminath Shauna, Maldives' Minister of Environment, Climate Change and Technology, who set out thoughts about the impacts of climate change on her country, and British Army General Richard Nugee, who told me of his experiences and insights linking climate change to armed conflict. I was very pleased to have the opportunity to speak with British philanthropist and ecological innovator Julia Davies and with Morris Pearl of Patriotic Millionaires USA. I was also delighted to speak with campaigner and writer Vandana Shiva, about justice, colonialism and Indigenous societies.

I also had the pleasure to speak with several Indigenous leaders. Komeok Joe leads the Penan organisation KERUAN and Celine Lim leads resistance for the Kayan people via their grassroots organisation SAVE Rivers, with both working to protect their ancestral rainforests in north Borneo. Domingo Peas is a leader and advocate for the Achuar people from the rainforests of Northern

Ecuador. Eli Enns is a First Nations Canadian from the Tla-o-qui-aht First Nation of Vancouver Island, and President of the Canadian section of the World Conservation Union. I also spoke with Tatiana Degai, from the Itelmen people from the Kamchatka Peninsula in eastern Russia. I am indebted to all of them for the thoughts, stories and experiences they contributed.

I have also been able to draw on the rich published literature accumulated by the thousands of scientists, researchers and analysts who work so tirelessly to shed light on the many challenges that face humankind at the start of the 21st century. Their fact and data-based endeavours are more important than ever, not least in bringing a rational dimension to debates that in divided societies are increasingly shaped by ideology, conspiracy theories and populism. I would like to acknowledge my gratitude for their efforts in helping to shine the lights of evidence, data and analysis into many of the grey and contested areas of debate covered in this book.

Artificial Intelligence applications were not used in the production of this book, other than transcription software employed to convert interviews into text files. All quotes were shared with interviewees for approval.

Finally, I would like to thank my wife Sue Sparkes, for all her support, and without whom I wouldn't get much done at all.

Image Credits

Figure 1 - Since 1950, the global population more than triples, to 8.2 billion in 2025. From Science of a Changing Planet by Tony Juniper © Dorling Kindersley Limited, A Penguin Random House Company. Source: UN, Department of Economic and Social Affairs, Population Division (2013), World Population Prospects: 'Most populous countries, 2014 and 2050', 2014 World Population Data Sheet, Population Reference Bureau, http://prb.org; Revised data: World Bank: https://data.worldbank.org/indicator/SP.POP.TOTL?locations=BR-CN-IN-ID-US

Figure 2 - Tenfold expansion in the global economy since 1950. From Science of a Changing Planet by Tony Juniper © Dorling Kindersley Limited, A Penguin Random House Company. Source: Our World in Data, Max Roser (2013) – 'Economic Growth'. Published online at OurWorldInData.org. Retrieved from: https://ourworldindata.org/economic-growth [Online Resource], CC BY 4.0

Figure 3 - Energy use up fivefold since the 1950s. From Science of a Changing Planet by Tony Juniper © Dorling Kindersley Limited, A Penguin Random House Company. Source: Our World in Data, https://ourworldindata.org/energy-production-consumption, CC BY 4.0 40–41

Figure 4 - Tenfold rise in consumption of natural resources. From Science of a Changing Planet by Tony Juniper © Dorling Kindersley Limited, A Penguin Random House Company. Source: F. Krausmann et al., Growth in global materials use, GDP and population during the 20th century, Ecological Economics, Vol. 68, Issue 10, 2009, pp. 2696–2705, ISSN 0921-8009, https://doi.org/10.1016/j.ecolecon.2009.05.007

Figure 5 - Mass extinction of animals and plants gather momentum. From Science of a Changing Planet by Tony Juniper © Dorling Kindersley Limited, A Penguin Random House Company. Source: GLOBIO3: A Framework to Investigate Options for Reducing Global Terrestrial Biodiversity Loss, Ecosystems (2009), 12, pp374–390, Rob Alkenmade, Mark van Oorschot, Lera Miles, Christian Nellemann, Michel Bakkenes, and Ben ten Brink, http://www.globio.info; Accelerated modern human-induced species losses: Entering the sixth mass extinction, Gerardo Ceballos, Paul R. Ehrlich, Anthony D. Barnosky, Andrés García, Robert M. Pringle and Todd M. Palmer, Science Advances, 19 June 2015, http://advances.sciencemag.org; Defaunation in the Anthropocene, Science, 25 July 2014, Vol. 345. Ossie 6195, pp401–406, http://science.sciencemag.org

Figure 6 - Record concentrations of greenhouse gases in the atmosphere. From Science of a Changing Planet by Tony Juniper © Dorling Kindersley Limited, A Penguin Random House Company. Data from: NOAA, 'Trends in Atmospheric Carbon Dioxide', Dr. Pieter Tans, NOAA/GML (gml.noaa.gov/ccgg/trends/) and Dr. Ralph Keeling, Scripps Institution of Oceanography (scrippsco2.ucsd.edu/) 110–111

Figure 7 - CO_2 Levels in the Atmosphere (by UNFCCC COP). CO_2 levels in atmosphere data: https://ourworldindata.org/grapher/co2-long-term-concentration.

Figure 8 - A country's level of health, social and environmental problems is significantly and strongly associated with inequality. © The Equality Trust/Kate Pickett and Richard Wilkinson, July 2024

Figure 9 - Percentage of CO_2 emissions by world populations. © Oxfam. The material 'Extreme Carbon Inequality report – Oxfam – December 2015 – (fig. 1, p.4)' is adapted by the publisher with the permission of Oxfam, Oxfam House, John Smith Drive, Cowley, Oxford OX4 2JY UK www.oxfam.org.uk. Oxfam does not necessarily endorse any text or activities that accompany the materials, nor has it approved the adapted text

Figure 10 - Carbon footprints by group across the world 2019. © Chancel, L., Bothe, P., Voituriez, T. (2023) Climate Inequality Report 2023, World Inequality Lab Study 2023/1. Source: Chancel (2022), courtesy of Philip Bothe

Figure 11 - UK home insulation improvements have plummeted since 2012. © Guardian News & Media Ltd 2024. Source: Climate Change Committee, Progress in reducing emissions March 2021 report

Figure 12 - Lack of trust is significantly and strongly associated with income inequality. © The Equality Trust/Kate Pickett and Richard Wilkinson, July 2024

Figure 13 - Lifting one third of the world population over the $3.2/day poverty line would increase global emissions by 5 per cent. Global top 1 per cent emissions are close to 15 per cent. © Chancel, L., Bothe, P., Voituriez, T. (2023), Climate Inequality Report 2023, World Inequality Lab Study 2023/1. Source: Bruckner et al. (2022), courtesy of Philip Bothe

Figure 14 - Life expectancy levels off at higher levels of economic development. © Richard Wilkinson and Kate Pickett. Source: Pickett, K. and Wilkinson, R., The Inner Level: How More Equal Societies Reduce Stress, Restore Sanity and Improve Everyone's Well-Being, 2018, (London: Allen Lane). Life expectancy data: https://data.worldbank.org/indicator/SP.DYN.LE00.IN, GDP per capita data: https://data.worldbank.org/indicator/NY.GDP.PCAP.CD

Notes

CHAPTER 1

1. World Population by Year – Worldometer, https://www.worldometers.info/world-population/world-population-by-year/ [accessed 24 March 2023].
2. Bar-On, Y. M., Phillips, R. and Milo, R. (2018), 'The biomass distribution on Earth', *Proceedings of the National Academy of Sciences*, 115, 6506–11.
3. Calvin, K., Dasgupta, D., Krinner, G., Mukherji, A., Thorne, P. W., Trisos, C., Romero, J., Aldunce, P. et al. (2023), IPCC (2023) *Climate Change 2023: synthesis report. Contribution of Working Groups I, II and III to the Sixth Assessment Report of the Intergovernmental Panel on Climate Change* (IPCC) (Core Writing Team: H. Lee and J. Romero, eds). Geneva: IPCC. doi:10.59327/IPCC/AR6-9789291691647
4. National Oceanic and Atmospheric Administration, 'Carbon dioxide now more than 50% higher than pre-industrial levels'. https://www.noaa.gov/news-release/carbon-dioxide-now-more-than-50-higher-than-pre-industrial-levels [accessed 18 March 2024].
5. NASA, 'Ice Sheets: Vital Signs – Climate Change: Vital Signs of the Planet'. https://climate.nasa.gov/vital-signs/ice-sheets/ [accessed 24 March 2023].
6. Press Office (2024), 'WMO confirms that 2023 smashes global temperature record', *World Meteorological Organization*. https://wmo.int/news/media-centre/wmo-confirms-2023-smashes-global-temperature-record [accessed 19 March 2024].
7. Press Office (2024), '2023: The warmest year on record globally – Met Office', *Met Office*. https://www.metoffice.gov.uk/about-us/press-office/news/weather-and-climate/2024/2023-the-warmest-year-on-record-globally [accessed 19 March 2024].
8. WWF (2022), *Living Planet Report 2022 – Building a nature positive society*. Almond, R.E.A., Grooten, M., Juffe Bignoli, D. & Petersen, T. (Eds). WWF, Gland, Switzerland.
9. Intergovernmental Science-Policy Platform on Biodiversity and Ecosystem Services (IPBES) (2019), CBD COP 15 Global Biodiversity Framework, Zenodo, doi:10.5281/ZENODO.3831673.
10. Sanderman, J., Hengl, T. and Fiske, G. J. (2017), 'Soil carbon debt of 12,000 years of human land use', *Proceedings of the National Academy of Sciences*, 114, 9575–80.
11. Potts, S. G. (2016), 'The assessment report on pollinators, pollination and food production: Summary for policymakers', Bonn, Germany: Secretariat of the Intergovernmental Science-Policy Platform on Biodiversity and Ecosystem Services (IPBES).

12 Wit, M. de, Hoogzaad, J. and Daniels, C. von. (2020), The Circularity Gap Report 2020, Amsterdam: Circle Economy.
13 Organisation for Economic Co-operation and Development (OECD) (2022), *Global Plastics Outlook: Economic Drivers, Environmental Impacts and Policy Options*. Paris: OECD Publishing. https://doi.org/10.1787/de747aef-en.
14 Kotz, M., Levermann, A. and Wenz, L. (2024), 'The economic commitment of climate change', *Nature*, 628, 551–7. https://doi.org/10.1038/s41586-024-07219-0.
15 Costanza, R., de Groot, R., Sutton, P., van der Ploeg S., Anderson, S. J., Kubiszewski, I., Farber, S. and Turner, R. K. (2014), 'Changes in the global value of ecosystem services', *Global Environmental Change*, 26, 152–8.
16 Juniper, T. (2020), 'This pandemic is an environmental issue', *London Evening Standard*, https://www.standard.co.uk/comment/comment/this-pandemic-is-an-environmental-issue-a4434651.html#r3z-addoor [accessed 18 March 2024].
17 Gupta, S. K., Minocha, R., Thapa, P. J., Srivastava, M. and Dandekar, T. (2022), 'Role of the pangolin in origin of SARS-CoV-2: an evolutionary perspective', *International Journal of Molecular Sciences*, 23, 9115.
18 Carlson, C. J., Albery, G. F., Merow, C., Trisos, C. H., Zipfel, C. M., Eskew, E. A., Olival, K. J., Ross, N. and Bansal, S. (2022), 'Climate change increases cross-species viral transmission risk', *Nature*, 607, 555–62.
19 Intergovernmental Panel on Climate Change (IPCC) and Masson-Delmotte, V. (2018), 'Summary for policymakers', *Global Warming of 1.5 °C: An IPCC Special Report on the impacts of global warming of 1.5 °C above pre-industrial levels and related global greenhouse gas emission pathways, in the context of strengthening the global response to the threat of climate change, sustainable development, and efforts to eradicate poverty* (Cambridge and New York: Cambridge University Press, doi:10.1017/9781009157940).
20 IPCC and Masson-Delmotte (2018), 'Summary for policymakers'.
21 Calvin et al. (2023), IPCC (2023), Climate Change 2023: synthesis report.

CHAPTER 2

1 Department of the Environment (1990), *This Common Inheritance*, London: HMSO.
2 Brundtland, G. H., World Commission on Environment and Development (eds), (1987), *Our Common Future*, Oxford: Oxford University Press.
3 Union of Concerned Scientists (1992), 'World Scientists' Warning to Humanity', https://www.ucsusa.org/resources/1992-world-scientists-warning-humanity#ucs-report-downloads.
4 Secretariat of the United Nations Framework Convention on Climate Change (1992), United Nations Framework Convention on Climate Change. Bonn, Germany: United Nations. unfccc.int/resource/docs/convkp/conveng.pdf.
5 United Nations Environment Programme (UNEP) (1992), Convention on biological diversity, June 1992. https://wedocs.unep.org/20.500.11822/8340.
6 Bolin, B. et al, (1995), IPCC Second Assessment synthesis of scientific-technical information relevant to interpreting Article 2 of the UN Framework Convention on Climate Change, Geneva, IPCC.
7 Pachauri, R. K., Reisinger, A. and IPCC (2007), Climate Change 2007: synthesis report. Contribution of Working Groups I, II and III to the Fourth Assessment Report of the Intergovernmental Panel on Climate Change. Geneva, Switzerland.

8   IPCC and Masson-Delmotte, V. (2021), Climate Change 2021 – The Physical Science Basis: Working Group I Contribution to the Sixth Assessment Report of the Intergovernmental Panel on Climate Change. Cambridge and New York: IPCC. doi:10.1017/9781009157896.

9   Pörtner, H.-O., Roberts, D., Tignor, M., Poloczanska, E., Mintenbeck, K., Alegría, A., Craig, S., Langsdorf, S. et al. (2023), Climate Change 2022 – Impacts, Adaptation and Vulnerability: Working Group II Contribution to the Sixth Assessment Report of the Intergovernmental Panel on Climate Change. Cambridge and New York: IPCC. doi:10.1017/9781009325844.

10 UNEP, Secretariat of the Convention on Biological Diversity, (2010), Strategic Plan for Biodiversity 2011 – 2020 including Aichi Biodiversity Targets, UNEP, CBD. https://www.cbd.int/sp/targets

11 United Nations (2014), Forests: Action Statements and Action Plans, The New York Declaration on Forests (Section 1). New York: United Nations.

12 The 17 Goals: Sustainable Development. https://sdgs.un.org/goals [accessed 12 March 2024].

13 Friedlingstein, P., O'Sullivan, M., Jones, M.W., Andrew, R. M., Gregor, L., Hauck, J., Le Quéré, C., Luijkx, I. T. et al. (2022), Global Carbon Budget 2022, *Earth System Science Data*, 14, 4811–900.

14 Emissions Gap Report 2022: The Closing Window – climate crisis calls for rapid transformation of societies (2022), Nairobi: UNEP. https://www.unep.org/emissions-gap-report-2022.

15 Convention on Biological Diversity (2020), Global Biodiversity Outlook 5: Summary for Policy Makers. Montreal: Secretariat of the Convention on Biological Diversity.

16 Schulte, I., Streck, C. and Roe, S. (2019), Protecting and Restoring Forests: a story of large commitments yet limited progress. New York Declaration on Forests Five-Year Assessment Report. forestdeclaration.org.

17 United Nations (2015), The Millennium Development Goals Report 2015: Summary. New York: United Nations.

18 Independent Group of Scientists appointed by the Secretary-General (2019), Global Sustainable Development Report 2019: The Future is Now – Science for Achieving Sustainable Development. New York: United Nations.

19 *BBC News* (2010), 'Biodiversity – a kind of washing powder?' https://www.bbc.com/news/science-environment-11546289 [accessed 19 March 2024].

CHAPTER 3

1   Szreter, S. (2021), *The History of Inequality*. Institute for Fiscal Studies.
2   Szreter, (2021), *History of Inequality*.
3   Desilver, D. (2013), 'US income inequality, on rise for decades, is now highest since 1928', Pew Research Center. https://www.pewresearch.org/short-reads/2013/12/05/u-s-income-inequality-on-rise-for-decades-is-now-highest-since-1928/ [accessed 5 April 2024].
4   Sommeiller, E., Price, M. and Wazeter E. (2016), 'Income inequality in the US by state, metropolitan area, and county', Economic Policy Institute.
5   'Gini coefficient by country' (2024), *World Population Review*. https://worldpopulationreview.com/country-rankings/gini-coefficient-by-country [accessed 12 March 2024].
6   World Bank (2022), 'Poverty and shared prosperity 2022: correcting course', doi:10.1596/978-1-4648-1893-6.

7   UNEP (2022) 'Global Multidimensional Poverty Index 2022: unpacking deprivation bundles to reduce multidimensional poverty', UNEP and Oxford Poverty and Human Development Initiative.
8   Chancel, L., Alvaredo, F., Piketty, T., Saez, E. and Zucman, G. (2018), World Inequality Report 2018: executive summary. Paris: World Inequality Lab.
9   Coffey, C., Espinoza Revollo, P., Harvey, R., Lawson, M., Parvez Butt, A., Piaget, K., Sarosi, D. and Thekkudan, J. (2020), 'Time to care: unpaid and underpaid care work and the global inequality crisis', Oxfam. doi:10.21201/2020.5419.
10 United Nations (2022) *Sustainable Development Goals Report 2022*. New York: United Nations. ISBN 978-92-1-101448-8.
11 Chancel, L., Piketty, T., Saez, E. and Zucman, G. (2022), World Inequality Report 2022. World Inequality Lab. wir2022.wid.world.
12 Kroll, C. (2015), *Sustainable Development Goals: Are the Rich Countries Ready?* Gutersloh: Bertelsmann Stiftung.
13 World Bank (ed.) (2020), *Poverty and Shared Prosperity 2020: Reversals of Fortune*. Washington, DC: World Bank.
14 Ahmed, N., Marriott, A., Dabi, N., Lowthers, M., Lawson, M. and Mugehera, L. (2022), 'Inequality kills: the unparalleled action needed to combat unprecedented inequality in the wake of COVID-19', Oxfam. doi:10.21201/2022.8465.
15 World Bank (2022), *Poverty and Shared Prosperity 2020*.
16 Chancel et al. (2022), World Inequality Report 2022.
17 Wilkinson, R. and Pickett, K. E. (2009), *The Spirit Level: Why More Equal Societies Almost Always Do Better*. London: Bloomsbury Publishing.
18 The Equality Trust, (2023). Cost of inequality 2023. https://equalitytrust.org.uk/evidence-base/reports/equality-trust-releases-cost-inequality-report/].
19 *BBC News* (2018) 'Timeline: how the BBC gender pay story has unfolded. https://www.bbc.co.uk/news/entertainment-arts-42833551 [accessed 19 March 2024].
20 Parliamentary Committees (2018), 'PAY0003: evidence on BBC pay', UK Parliamentary Committees. https://committees.parliament.uk/writtenevidence/86362/html/ [accessed 19 March 2024].
21 Reuters (2018), 'Claire Foy to get £200,000 after gender pay row on *The Crown*', *The Guardian*. https://www.theguardian.com/tv-and-radio/2018/may/01/claire-foy-to-get-200000-after-gender-pay-row-on-the-crown-reports [accessed 9 February 2023].
22 Flood, A. (2018), Books by women priced 45% lower, study finds. *The Guardian*. Available at: https://www.theguardian.com/books/2018/may/01/books-by-women-priced-45-lower-study-finds [accessed 6 March 2023].
23 Office for National Statistics (2022), 'Gender pay gap in the UK'. London: ONS.
24 Fawcett Society (2022), 'Equal pay day 2022: what does the gender pay gap mean for women in the cost-of-living crisis?' London: The Fawcett Society.
25 Chancel et al (2022), World Inequality Report 2022.
26 Neef, T. and Robilliard, A.-S. (2021), *Half the Sky? The Female Labor Income Share in a Global Perspective*. France: World Inequality Lab.
27 Dong, Y., Morgan, C., Chinenov, Y., Zhou, L., Fan, W., Ma, X. and Pechenkina, K. (2017), 'Shifting diets and the rise of male-biased inequality on the Central Plains of China during Eastern Zhou', *Proceedings of the National Academy of Sciences*, 114, 932–7.
28 Cintas-Peña, M. and García Sanjuán, L. (2019), 'Gender inequalities in Neolithic Iberia: a multi-proxy approach', *European Journal of Archaeology*, 22, 499–522.

29 Live Science (2025). 'Were the Celts matriarchal? Ancient DNA reveals men married into local, powerful female lineages', https://www.livescience.com/archaeology/were-the-celts-matriarchal-ancient-dna-reveals-men-married-into-local-powerful-female-lineages [accessed 18 January 2025].
30 Married Women's Property Act 1882 (1882). https://www.legislation.gov.uk/ukpga/Vict/45-46/75/introduction/enacted [accessed 26 March 2024].
31 Buchanan, I. (2024), 'Women in politics and public life'. London: House of Commons.
32 Fernandez-Armesto, F. (2013), *1492: The Year Our World Began*. London: Bloomsbury Publishing.
33 Gigoux, C. (2022). '"Condemned to disappear": Indigenous genocide in Tierra del Fuego', *Journal of Genocide Research*, 24, 1–22.
34 Lothrop, S. K. (1928), *The Indians of Tierra del Fuego*. New York: Museum of the American Indian, Heye Foundation.
35 Jalata, A. (2013), 'The impacts of English colonial terrorism and genocide on Indigenous/Black Australians', *SAGE Open*, 3, 215824401349914.
36 Jalata (2013), 'The impacts of English colonial terrorism'.
37 Victor-Pujebet, B. and Blanchard, P. (directors) (2018), *Inside Human Zoos*. France: Bonne Pioche & Archipel.
38 Půtová, B. (2018), 'Freak shows: otherness of the human body as a form of public presentation', *Anthropologie*, 56, 91–102.
39 Mülchi, H. (2011), *The Human Zoo: The Final Journey of Calafate*. New York: Icarus Films.
40 Boffey, D. (2018), 'Belgium comes to terms with "human zoos" of its colonial past', *The Guardian*. https://www.theguardian.com/world/2018/apr/16/belgium-comes-to-terms-with-human-zoos-of-its-colonial-past [accessed 15 March 2023].
41 National Museum of Australia (2022), 'End of the White Australia policy'. https://www.nma.gov.au/defining-moments/resources/end-of-white-australia-policy [accessed 28 March 2024].
42 Prasad, R. (2023), '"I can't breathe": race, death and British policing'. London: Inquest. https://www.inquest.org.uk/police-racism-report-2023.
43 Dey, M., White, C. and Kaur, S. (2021), *Pay and Progression of Women of Colour*. London: The Fawcett Society, The Runnymede Trust.
44 Parker Review Committee (2023), 'Improving the ethnic diversity of UK business: an update from the Parker Review'. The Parker Review Committee.
45 Marren, C. and Bazeley, A. (2022), 'Sex & power 2022'. London: The Fawcett Society.
46 Brodnock, E. (2020), *Diversity Beyond Gender: The State of the Nation for Diverse Entrepreneurs*. London: Extend Ventures.
47 Morales, D. R. and Ali, S. N. (2021), 'COVID-19 and disparities affecting ethnic minorities', *The Lancet*, 397, 1684–5.
48 Nazroo, J. and Bécares, L. (2021), 'Ethnic inequalities in COVID-19 mortality: a consequence of persistent racism', Runnymede/CoDE COVID Briefings, University of Manchester Research.
49 Currenti, R. and Flatley, J. (2020), 'Policing the pandemic: detailed analysis on police enforcement of the public health regulations and an assessment on disproportionality across ethnic groups', National Police Chiefs' Council.
50 Montacute, R. (2019), 'Elitist Britain 2019: the educational backgrounds of Britain's leading people'. London: Sutton Trust, Social Mobility Commission.
51 The Birthstrike Movement (2024), https://birthstrikemovement.org/ [accessed 5 April 2024].

NOTES

CHAPTER 4

1. Perlin, S. A., Sexton, K. and Wong, D. W. S. (1999), 'An examination of race and poverty for populations living near industrial sources of air pollution', *Journal of Exposure Science and Environmental Epidemiology*, 9, 29–48.
2. Environmental Protection Agency (EPA) (2020), Annual Environmental Justice Progress Report FY2020. USA: Environmental Protection Agency. https://www.epa.gov/environmentaljustice/annual-environmental-justice-progress-reports.
3. Massey, R. (2004), *Environmental Justice: Income, Race, and Health*. Medford, MA: Global Development and Environment Institute, Tufts University.
4. Cabrera, Y., Smith Hopkins, J. and Moran, G. (2023), 'EPA promised to address environmental racism. Then states pushed back', Center for Public Integrity. https://publicintegrity.org/environment/pollution/environmental-justice-denied/environmental-justice-epa-civil-rights-story/ [accessed 29 April 2024].
5. Fulwood III, S. (2016), 'The United States' history of segregated housing continues to limit affordable housing', Center for American Progress. https://www.americanprogress.org/article/the-united-states-history-of-segregated-housing-continues-to-limit-affordable-housing/ [accessed 29 April 2024].
6. Gross T. (2017), 'A "forgotten history" of how the US government segregated America', National Public Radio (NPR). https://www.npr.org/2017/05/03/526655831/a-forgotten-history-of-how-the-u-s-government-segregated-america [accessed 29 April 2024].
7. Lee, C. (1987), 'Toxic wastes and race in the United States: a national report of racial and socio-economic characteristics of communities with hazardous waste sites', Commission for Racial Justice, United Church of Christ.
8. 'Chemical complex checks by residents' (2003), *The Northern Echo*. https://www.thenorthernecho.co.uk/news/7013969.chemical-complex-checks-residents/ [accessed 28 April 2024].
9. Friends of the Earth (FOE) et al. (2003), 'Behind the shine: the other Shell report 2003', London: FOE, Coletivo Alternative Verde (CAVE), Community In-power Development Association (CIDA), Concerned Citizens of Norco, Environmental Rights Action of Nigeria (FOE Nigeria), Global Community Monitor (GCM), groundWork (FOE South Africa) & groundWork USA, Louisiana Bucket Brigade, Sakhalin Environmental Watch, South Durban Community Environmental Alliance (SDCEA), and United Front to Oust Oil Depots (UFO-OD).
10. Whitty C. (2022), Chief Medical Officer's Annual Report 2022: air pollution. https://assets.publishing.service.gov.uk/media/6389ee858fa8f569f9c823d2/executive-summary-and-recommendations-air-pollution.pdf
11. World Health Organization (2022), 'Ambient (outdoor) air pollution', WHO. https://www.who.int/news-room/fact-sheets/detail/ambient-(outdoor)-air-quality-and-health [accessed 8 April 2024].
12. Whitty (2022), Chief Medical Officer's Annual Report 2022.
13. Friends of the Earth (2022), 'Which neighbourhoods have the worst air pollution?' Friends of the Earth Policy and Insight. https://policy.friendsoftheearth.uk/insight/which-neighbourhoods-have-worst-air-pollution [accessed 15 February 2023].
14. *BBC News* (2020), 'Ella Adoo-Kissi-Debrah: air pollution a factor in girl's death, inquest finds'. https://www.bbc.com/news/uk-england-london-55330945 [accessed 8 April 2024].

15 The Ella Roberta Foundation Ella's Law (2022). https://www.ellaroberta.org/campaigns/ellas-law [accessed 8 April 2024].
16 London Assembly (2022), 'Assembly supports "Ella's Law"', London City Hall. https://www.london.gov.uk/who-we-are/what-london-assembly-does/london-assembly-press-releases/assembly-supports-ellas-law [accessed 8 April 2024].
17 Scheidel, A., Fernández-Llamazares, Á., Bara, A. H., Del Bene, D., David-Chavez, D. M., Fanari, E., Garba, I., Hanaček, K. et al. (2023), 'Global impacts of extractive and industrial development projects on Indigenous Peoples' lifeways, lands, and rights', *Science Advances*, 9(23). doi: 10.1126/sciadv.ade9557
18 Ninomiya, M. E. M., Burns, N., Pollock, N. J., Green, N. T. G., Martin, J., Linton, J., Rand, J. R., Brubacher, L. J. et al. (2023), 'Indigenous communities and the mental health impacts of land dispossession related to industrial resource development: a systematic review', *The Lancet Planetary Health*, 7, e501–17.
19 Leyton-Flor, S. A. and Sangha, K. (2024), 'The socio-ecological impacts of mining on the well-being of Indigenous Australians: a systematic review', *The Extractive Industries and Society*, 17, 101429.
20 The Borneo Project and Manser Fonds, Bruno (2023), 'Lost in certification: how forest certification greenwashes Samling's dirty timber and fools the international market'. https://bit.ly/lostincert.
21 SAVE Rivers, 'Sign the petition to stop the certification of conflict timber in Sarawak'. https://saverivers.org/stopthechop/ [accessed 8 April 2024].
22 The Borneo Project (2023), 'Stop the SLAPP'. https://borneoproject.org/tag/stop-the-slapp/ [accessed 29 April 2024].
23 Amazon Sacred Headwaters Alliance. https://sacredheadwaters.org/ [accessed 8 April 2024].
24 United Kingdom Without Incineration Network (UKWIN). 'UK incinerators'. https://ukwin.org.uk/incinerators/?filter=&region=&potential=show [accessed 8 April 2024].
25 Hooper, R. (2012), 'Villagers lose St Dennis waste plant battle', *The Independent*. https://www.independent.co.uk/news/uk/home-news/villagers-lose-st-dennis-waste-plant-battle-8005177.html [accessed 29 April 2024].
26 Trewhela, L. (2022), 'Village in Cornwall's Clay Country may be deprived but its residents wouldn't live anywhere else', *Cornwall Live*. https://www.cornwalllive.com/news/cornwall-news/village-cornwalls-clay-country-deprived-7847198 [accessed 29 April 2024].
27 Consumer Data Research Centre (2024). Index of Multiple Deprivation (IMD), CDRC data. https://data.cdrc.ac.uk/dataset/index-multiple-deprivation-imd [accessed 8 April 2024].
28 Roy, I. (2020), 'Waste incinerators three times more likely to be sited in UK's most deprived neighbourhoods', Unearthed. https://unearthed.greenpeace.org/2020/07/31/waste-incinerators-deprivation-map-recycling/ [accessed 25 February 2023].
29 Karlsson, T., Dell, J., Gündoğdu, S. and Carney Almroth, B. (2023), 'Plastic waste trade: the hidden numbers', International Pollutants Elimination Network (IPEN). https://ipen.org/documents/plastic-waste-trade-hidden-numbers [accessed 22 March 2023].
30 Organisation for Economic Co-operation and Development (OECD) (2022), *Global Plastics Outlook: Economic Drivers, Environmental Impacts and Policy Options*. Paris: OECD Publishing. https://doi.org/10.1787/de747aef-en.

31 Environment, Food and Rural Affairs Committee (2022), 'The price of plastic: ending the toll of plastic waste.' Report by the Environment, Food and Rural Affairs Committee to Parliament. London: House of Commons.
32 Parliament Committees (2021), 'Written evidence submitted by the British Plastic Federation (PW0042)'. Parliament Committees. https://committees.parliament.uk/writtenevidence/38902/html/ [accessed 8 April 2024].
33 Department for Environment, Food and Rural Affairs (2023), UK statistics on waste: GOVUK. https://www.gov.uk/government/statistics/uk-waste-data/uk-statistics-on-waste#packaging-waste [accessed 19 February 2024].
34 Cobbing, M., Daaji, S., Kopp, M., Wohlgemuth, V., Dena, H. K., Dreher, T., Hüttemann, A. and Zils, M. (2022), Report: Poisoned Gifts. From donations to the dumpsite: textile waste disguised as second-hand clothes exported to East Africa. Greenpeace Germany.
35 Besser, L. (2021), 'Dead white man's clothes: how fast fashion is turning parts of Ghana into toxic landfill', *ABC News*. https://www.abc.net.au/news/2021-08-12/fast-fashion-turning-parts-ghana-into-toxic-landfill/100358702 [accessed 19 February 2024].
36 Karlsson et al. (2023), 'Plastic waste trade'.
37 Petrlik, J., Ismawati, Y., DiGangi, J., Arisandi, P., Si, M., Bell, L. and Beeler, B. (2019), 'Plastic waste flooding Indonesia leads to toxic chemical contamination of the food chain', IPEN. https://ipen.org/news/plastic-waste-poisons-indonesia%E2%80%99s-food-chain.
38 OECD (2022), *Global Plastics Outlook*.
39 EPA (2020), Annual Environmental Justice Progress Report FY2020.

CHAPTER 5

1 US Department of State (2023), Investment Climate Statements: Maldives. United States Department of State – Maldives. https://www.state.gov/reports/2023-investment-climate-statements/maldives/ [accessed 15 January 2024].
2 UNEP (2021), 'Making peace with nature: a scientific blueprint to tackle the climate, biodiversity and pollution crisis'. United Nations Environment Programme. https://www.unep.org/resources/making-peace-nature.
3 Chakrabarti, S., Scott, S. P., Alderman, H., Menon, P. and Gilligan, D. O. (2021), 'Intergenerational nutrition benefits of India's national school feeding program', *Nature Communications*, 12, 4248.
4 Chakravarty, N., Tatwadi, K. and Ravi, K. (2019), 'Intergenerational effects of stunting on human capital: where does the compass point?' *International Journal of Medicine and Public Health*, 9, 105–11.
5 Cooper, M. W., Brown, M. E., Hochrainer-Stigler, S., Pflug, G., McCallum, I., Fritz, S., Silva, J. and Zvoleff, A. (2019), 'Mapping the effects of drought on child stunting', *Proceedings of the National Academy of Sciences*, 116, 17219–24.
6 United Nations (2023), 'In Mogadishu, UN chief urges "massive international support" for Somalia', *United Nations News*. https://news.un.org/en/story/2023/04/1135492 [accessed 17 April 2023].
7 Hufstader, C. (2022), 'Threat of "imminent" famine in Somalia', Oxfam. https://www.oxfamamerica.org/explore/stories/famine-in-somalia/ [accessed 15 March 2023].

8   Lowder, S. K., Sánchez, M. V. and Bertini, R. (2021), 'Which farms feed the world and has farmland become more concentrated?' *World Development*, 142, 105455.
9   Grainger-Jones, E. (2011) 'Climate change: building smallholder resilience', International Fund for Agricultural Development. https://www.ifad.org/en/issues [accessed 22 April 2024].
10 FAO, AUC, ECA and WFP, (2023), Africa: regional overview of food security and nutrition 2023, Statistics and trend' (Accra, FAO, AUC, United Nations Economic Commission for Africa (ECA), WFP. doi:10.4060/cc8743en.
11 FAO, IFAD, UNICEF, WFP and WHO, 'The State of Food Security and Nutrition in the World 2023. Urbanization, agrifood systems transformation and healthy diets across the rural–urban continuum. Rome, FAO,. doi:10.4060/cc3017en.
12 IPPC Secretariat, (2021), 'Scientific review of the impact of climate change on plant pests'. A global challenge to prevent and mitigate plant pest risks in agriculture, forestry and ecosystems, Rome, FAO on behalf of the IPPC Secretariat. doi:10.4060/cb4769en.
13 Heino, M., Kinnunen, P., Anderson, W., Ray, D. K., Puma, M. J., Varis, O., Siebert, S. and Kummu, M. (2023), 'Increased probability of hot and dry weather extremes during the growing season threatens global crop yields', *Scientific Reports*, 13, 3583.
14 Ratcliff, A. (2015), 'Hot and hungry: how to stop climate change derailing the fight against hunger', Oxfam International. https://primarysources.brillonline.com/browse/human-rights-documents-online/hot-and-hungry-how-to-stop-climate-change-derailing-the-fight-against-hunger;hrdhrd98242014012 [accessed 9 April 2024].
15 'Record number of fossil fuel lobbyists at COP28: press release' (2023), Kick Big Polluters Out Coalition. https://kickbigpollutersout.org/articles/release-record-number-fossil-fuel-lobbyists-attend-cop28 [accessed 9 April 2024].
16 Alsalem, R. (2022), 'Violence against women and girls, its causes and consequences', United Nations General Assembly.
17 Office of the High Commissioner for Human Rights (OHCHR) (2022), 'Climate change exacerbates violence against women and girls', United Nations Human Rights Office of the High Commissioner. https://www.ohchr.org/en/stories/2022/07/climate-change-exacerbates-violence-against-women-and-girls [accessed 9 April 2024].
18 Erman, A., De Vries Robbe, S. A., Thies, S. F., Kabir, K. and Maruo, M. (2021), *Gender Dimensions of Disaster Risk and Resilience: Existing Evidence*. World Bank. doi:10.1596/35202.
19 Reyes, R. R. (2002), 'Gendering responses to El Niño in rural Peru', *Gender and Development*, 10.
20 Pattisson, P., McIntyre, N. and Mukhtar, I. (2021), 'Revealed: 6,500 migrant workers have died in Qatar since World Cup awarded', *The Guardian*. https://www.theguardian.com/global-development/2021/feb/23/revealed-migrant-worker-deaths-qatar-fifa-world-cup-2022 [accessed 13 April 2023].
21 Nerbass, F. B., Pecoits-Filho, R., Clark, W. F., Sontrop, J. M., McIntyre, C. W. and Moist, L. (2017), 'Occupational heat stress and kidney health: from farms to factories,' *Kidney International Reports*, 2, 998–1008.
22 Ansah, E. W., Ankomah-Appiah, E., Amoadu, M. and Sarfo, J. O. (2021), 'Climate change, health and safety of workers in developing economies: a scoping review', *Journal of Climate Change and Health*, 3, 100034.

23 Tuholske, C., Caylor, K., Funk, C., Verdin, A., Sweeney, S., Grace, K., Peterson, P. and Evans, T. (2021), 'Global urban population exposure to extreme heat', *Proceedings of the National Academy of Sciences*, 118, e2024792118.
24 Jung, M. W., Haddad, M. A. and Gelder, B. K. (2024), 'Examining heat inequity in a Brazilian metropolitan region', *Environment and Planning B: Urban Analytics and City Science*, 51, 109–27.
25 ONS and UKHSA (2022), 'Excess mortality during heat-periods: 1 June to 31 August 2022', Office for National Statistics and UK Health Security Agency. https://www.ons.gov.uk/peoplepopulationandcommunity/birthsdeathsandmarriages/deaths/articles/excessmortalityduringheatperiods/englandandwales1juneto31august2022.
26 Hsu, A., Sheriff, G., Chakraborty, T. and Manya, D. (2021), 'Disproportionate exposure to urban heat island intensity across major US cities', *Nature Communications*, 12, 2721.
27 Lindley, S. (2022), 'Communities most vulnerable to climate change: background briefing', University of Manchester & Friends of the Earth UK. https://friendsoftheearth.uk/climate/millions-risk-extreme-heat-unless-climate-goals-met [accessed 9 April 2024].
28 Dialesandro, J., Brazil, N., Wheeler, S. and Abunnasr, Y. (2021), 'Dimensions of thermal inequity: neighborhood social demographics and urban heat in the southwestern US', *International Journal of Environmental Research and Public Health*, 18, 941.
29 Bradshaw, H., England, R. and Finnerty, D. (2022), '"It's like an oven": life in Britain's hottest neighbourhoods', *BBC News*. https://www.bbc.co.uk/news/uk-62126463 [accessed 29 April 2024].
30 Doherty, C. (2015), 'Remembering Katrina: wide racial divide over government's response', Pew Research Center. https://www.pewresearch.org/fact-tank/2015/08/27/remembering-katrina-wide-racial-divide-over-governments-response/ [accessed 13 April 2023].
31 Kelley, C. P., Mohtadi, S., Cane, M. A., Seager, R. and Kushnir, Y. (2015), 'Climate change in the Fertile Crescent and implications of the recent Syrian drought', *Proceedings of the National Academy of Sciences*, 112, 3241–6.
32 Olumba, E. E., Nwosu, B. U., Okpaleke, F. N. and Okoli, R. C. (2022), 'Conceptualising eco-violence: moving beyond the multiple labelling of water and agricultural resource conflicts in the Sahel', *Third World Quarterly*, 43, 2075–90.
33 Bilak, A. (2016), Global Report on Internal Displacement 2016. Switzerland: Internal Displacement Monitoring Centre (IDMC). https://www.internal-displacement.org/globalreport2016/#home.
34 Bilak, A. (2023), Global Report on Internal Displacement 2023: Internal Displacement and Food Security. Switzerland: IDMC. https://www.internal-displacement.org/global-report/grid2023/.

CHAPTER 6

1 Guts UK (2019), 'The role of gut bacteria in health and disease', Guts UK and British Society of Gastroenterology. https://gutscharity.org.uk/advice-and-information/health-and-lifestyle/the-role-of-gut-bacteria-in-health-and-disease/.
2 Editorial Contributors (2022), 'Your gut bacteria and your health', WebMD. https://www.webmd.com/digestive-disorders/what-your-gut-bacteria-say-your-health [accessed 10 May 2024].

3   Wilson, Edward O. (1986), *Biophilia*. Cambridge, MA: Harvard University Press.
4   UN (2019), 'World urbanization prospects, 2018: highlights', New York: United Nations.
5   Marselle, M. R., Lindley, S. J., Cook, P. A. and Bonn, A. (2021), 'Biodiversity and health in the urban environment', *Current Environmental Health Reports*, 8, 146–56.
6   Aerts, R., Honnay, O. and Van Nieuwenhuyse, A. (2018), 'Biodiversity and human health: mechanisms and evidence of the positive health effects of diversity in nature and green spaces', *British Medical Bulletin*, 127, 5–22.
7   Berto, R. and Barbiero, G. (2017), 'How the psychological benefits associated with exposure to nature can affect pro-environmental behavior', *Annals of Cognitive Science*, 1, 16–20.
8   Astell-Burt, T. and Feng, X. (2019), 'Urban green space, tree canopy, and prevention of heart disease, hypertension, and diabetes: a longitudinal study', *The Lancet Planetary Health*, 3, S16.
9   Meo, S. A., Halepoto, D. M., Meo, A. S. and Klonoff, D. C. (2022), 'Impact of green space environment on the prevalence of diabetes mellitus in European countries', *Journal of King Saud University – Science*, 34, 102269.
10 Ccami-Bernal, F., Soriano-Moreno, D. R., Fernandez-Guzman, D., Tuco, K. G., Castro-Díaz, S. D., Esparza-Varas, A. L., Medina-Ramirez, S. A., Caira-Chuquineyra, B. et al. (2023), 'Green space exposure and type 2 diabetes mellitus incidence: a systematic review', *Health & Place*, 82, 103045.
11 Dadvand, P., Villanueva, C. M., Font-Ribera, L., Martinez, D., Basagaña, X., Belmonte, J., Vrijheid, M., Gražulevičienė, R. et al. (2014), 'Risks and benefits of green spaces for children: a cross-sectional study of associations with sedentary behavior, obesity, asthma, and allergy', *Environmental Health Perspectives*, 122, 1329–35.
12 Banay, R., Bezold, C., James, P., Hart, J. and Laden, F. (2017), 'Residential greenness: current perspectives on its impact on maternal health and pregnancy outcomes', *International Journal of Women's Health*, 9, 133–44.
13 Dadvand et al. (2014), 'Risks and benefits of green spaces'.
14 Engemann, K., Pedersen, C. B., Arge, L., Tsirogiannis, C., Mortensen, P. B. and Svenning, J.-C. (2018), 'Childhood exposure to green space – a novel risk-decreasing mechanism for schizophrenia?' *Schizophrenia Research*, 199, 142–8.
15 Ruokolainen, L., Hertzen, L., Fyhrquist, N., Laatikainen, T., Lehtomäki, J., Auvinen, P., Karvonen, A. M., Hyvärinen, A. et al. (2015), 'Green areas around homes reduce atopic sensitization in children', *Allergy*, 70, 195–202.
16 Bijnens, E. M., Nawrot, T. S., Loos, R. J., Gielen, M., Vlietinck, R., Derom, C. and Zeegers, M. P. (2017), 'Blood pressure in young adulthood and residential greenness in the early-life environment of twins', *Environmental Health*, 16, 53.
17 Aerts et al. (2018), 'Biodiversity and human health'.
18 Vella-Brodrick, D. A. and Gilowska, K. (2022), 'Effects of nature (greenspace) on cognitive functioning in school children and adolescents: a systematic review', *Educational Psychology Review*, 34, 1217–54.
19 Moran, D. and Turner, J. (2019), 'Turning over a new leaf: the health-enabling capacities of nature contact in prison', *Social Science & Medicine*, 231, 62–9.
20 Ulrich, R. S. (1984), 'View through a window may influence recovery from surgery', *Science*, 224, 420–1.

21 Manes, F., Marando, F., Capotorti, G., Blasi, C., Salvatori, E., Fusaro, L., Ciancarella, L., Mircea, M. et al. (2016), 'Regulating ecosystem services of forests in ten Italian metropolitan cities: air quality improvement by PM10 and O3 removal', *Ecological Indicators*, 67, 425–40.
22 Zhang, Y.-D., Zhou, G.-L., Wang, L., Browning, M. H. E. M., Markevych, I., Heinrich, J., Knibbs, L. D., Zhao, T. et al. (2024), 'Greenspace and human microbiota: a systematic review', *Environment International*, 187, 108662.
23 Ege, M. J. (2011), 'Exposure to environmental microorganisms and childhood asthma', *New England Journal of Medicine*, 364, 8, 701–09.
24 Marselle et al. (2021), 'Biodiversity and health'.
25 Cameron, R. W. F., Brindley, P., Mears, M., McEwan, K., Ferguson, F., Sheffield, D., Jorgensen, A., Riley, J. et al. (2020), 'Where the wild things are! Do urban green spaces with greater avian biodiversity promote more positive emotions in humans?' *Urban Ecosystems*, 23, 301–17.
26 Kuo, F. E. and Sullivan, W. C. (2001), 'Aggression and violence in the inner city: effects of environment via mental fatigue', *Environment and Behavior*, 33, 543–71.
27 Ewane, E. B., Bajaj, S., Velasquez-Camacho, L., Srinivasan, S., Maeng, J., Singla, A., Luber, A., de-Miguel, S. et al. (2023), 'Influence of urban forests on residential property values: a systematic review of remote sensing-based studies', *Heliyon*, 9, e20408.
28 Ogletree, S. S., Larson, L. R., Powell, R. B., White, D. L. and Brownlee, M. T. J. (2022), 'Urban greenspace linked to lower crime risk across 301 major US cities', *Cities*, 131, 103949.
29 Martin, L., White, M. P., Hunt, A., Richardson, M., Pahl, S. and Burt, J. (2020), 'Nature contact, nature connectedness and associations with health, wellbeing and pro-environmental behaviours', *Journal of Environmental Psychology*, 68, 101389.
30 Martin et al. (2020), 'Nature contact, nature connectedness'.
31 Alcock, I., White, M. P., Pahl, S., Duarte-Davidson, R. and Fleming, L. E. (2020), 'Associations between pro-environmental behaviour and neighbourhood nature, nature visit frequency and nature appreciation: evidence from a nationally representative survey in England', *Environment International*, 136, 105441.
32 Rosa, C. D., Profice, C. C. and Collado, S. (2018), 'Nature experiences and adults' self-reported pro-environmental behaviors: the role of connectedness to nature and childhood nature experiences', *Frontiers in Psychology*, 9, 1055.
33 Molinario, E., Lorenzi, C., Bartoccioni, F., Perucchini, P., Bobeth, S., Colléony, A., Diniz, R., Eklund, A. et al. (2020), 'From childhood nature experiences to adult pro-environmental behaviors: an explanatory model of sustainable food consumption', *Environmental Education Research*, 26, 1137–63.
34 Collado, S. and Evans, G. W. (2019), 'Outcome expectancy: a key factor to understanding childhood exposure to nature and children's pro-environmental behavior', *Journal of Environmental Psychology*, 61, 30–6.
35 Charles, C., Keenleyside, K., Chapple, R. and Kilburn, B. (2018), 'Home to us all: how connecting with nature helps us care for ourselves and the Earth', Children & Nature Network (USA), International Union for Conservation of Nature (IUCN). https://www.iucn.org/resources/grey-literature/home-us-all-how-connecting-nature-helps-us-care-ourselves-and-earth.
36 Molinario et al. (2020), 'From childhood nature experiences'.

37 Otto, S. and Pensini, P. (2017), 'Nature-based environmental education of children: environmental knowledge and connectedness to nature, together, are related to ecological behaviour', *Global Environmental Change*, 47, 88–94.
38 Collado, S., Staats, H. and Corraliza, J. A. (2013), 'Experiencing nature in children's summer camps: affective, cognitive and behavioural consequences', *Journal of Environmental Psychology*, 33, 37–44.
39 Charles et al. (2018), 'Home to us all'.
40 Rosa et al. (2018), 'Nature experiences and adults' self-reported pro-environmental behaviors'.
41 Alcock et al. (2020), 'Associations between pro-environmental behaviour'.
42 Piaggio, M. (2021), 'The value of public urban green spaces: measuring the effects of proximity to and size of urban green spaces on housing market values in San José, Costa Rica', *Land Use Policy*, 109, 105656.
43 Trojanek, R., Gluszak, M. and Tanas, J. (2018), 'The effect of urban green spaces on house prices in Warsaw', *International Journal of Strategic Property Management*, 22, 358–71.
44 Ma, Y., Koomen, E., Rouwendal, J. and Wang, Z. (2024), 'The increasing value of urban parks in a growing metropole', *Cities*, 147, 104794.
45 Glover, J. (2019), Landscape Review: Final Report. London: Department for Environment, Food and Rural Affairs. https://www.gov.uk/government/publications/designated-landscapes-national-parks-and-aonbs-2018-review/landscapes-review-summary-of-findings.
46 European Environment Agency (2022), 'Who benefits from nature in cities? Social inequalities in access to urban green and blue spaces across Europe'. https://www.eea.europa.eu/publications/who-benefits-from-nature-in/folder_contents.
47 Rehling, J., Bunge, C., Waldhauer, J. and Conrad, A. (2021), 'Socioeconomic differences in walking time of children and adolescents to public green spaces in urban areas – results of the German Environmental Survey (2014–2017)', *International Journal of Environmental Research and Public Health*, 18, 2326.
48 Trojanek et al. (2018), 'The effect of urban green spaces on house prices'.
49 Csomós, G., Farkas, J. Z. and Kovács, Z. (2020), 'Access to urban green spaces and environmental inequality in post-socialist cities', *Hungarian Geographical Bulletin*, 69, 191–207.
50 Vierikko, K., Gonçalves, P., Haase, D., Elands, B., Ioja, C., Jaatsi, M., Pieniniemi, M., Lindgren, J. et al. (2020), 'Biocultural diversity (BCD) in European cities – interactions between motivations, experiences and environment in public parks', *Urban Forestry & Urban Greening*, 48, 126501.
51 De Vries, S., Buijs, A. E. and Snep, R. P. H. (2020), 'Environmental justice in the Netherlands: presence and quality of greenspace differ by socioeconomic status of neighbourhoods', *Sustainability*, 12, 5889.
52 Hoffimann, E., Barros, H. and Ribeiro, A. (2017), 'Socioeconomic inequalities in green space quality and accessibility—evidence from a southern European city', *International Journal of Environmental Research and Public Health*, 14, 916.
53 Commission for Architecture and the Built Environment (2010), 'Community green: using local spaces to tackle inequality and improve health'. London: Commission for Architecture and the Built Environment.

54 Goldenberg, R., Kalantari, Z. and Destouni, G. (2018), 'Increased access to nearby green–blue areas associated with greater metropolitan population well-being', *Land Degradation & Development*, 29, 3607–16.
55 Łaszkiewicz, E. and Sikorska, D. (2020), 'Children's green walk to school: an evaluation of welfare-related disparities in the visibility of greenery among children', *Environmental Science & Policy*, 110, 1–13.
56 Astell-Burt and Feng (2019), 'Urban green space'.
57 Williams, T. G., Logan, T. M., Zuo, C. T., Liberman, K. D. and Guikema, S. D. (2020), 'Parks and safety: a comparative study of green space access and inequity in five US cities', *Landscape and Urban Planning*, 201, 103841.
58 Rigolon, A., Browning, M., Lee, K. and Shin, S. (2018), 'Access to urban green space in cities of the Global South: a systematic literature review', *Urban Science*, 2, 67.
59 de Zylva, P., Gordon-Smith, C. and Childs, M. (2020), 'England's green space gap: how to end green space deprivation in England', Friends of the Earth. https://policy.friendsoftheearth.uk/insight/englands-green-space-gap.
60 IFF Research (2023), 'People and Nature Survey analysis: findings report', Natural England. https://publications.naturalengland.org.uk/.
61 The Ramblers Association (2020), 'The grass isn't greener for everyone: why access to green space matters', The Ramblers Association. https://www.ramblers.org.uk/news/features.
62 Commission for Architecture and the Built Environment (2010), 'Community green'.
63 Natural England (2022), Green Infrastructure Mapping Database and Analyses – Version 1.2. https://designatedsites.naturalengland.org.uk/GreenInfrastructure/MappingAnalysis.aspx [accessed 6 May 2024].
64 Lacobucci, G. (2021), 'One in 10 UK adults could have diabetes by 2030, warns charity', *The BMJ: British Medical Journal*, n2453.
65 Diabetes UK (2024), 'Type 2 Diabetes Prevention Week 2024: action needed to tackle rising type 2 cases among under-40s'. https://www.diabetes.org.uk/about-us/news-and-views/diabetes-prevention-week-2024-action-needed-tackle-rising-type-2-cases [accessed 26 May 2024].
66 Cockerham, W. C. (2022), 'Theoretical approaches to research on the social determinants of obesity', *American Journal of Preventive Medicine*, 63, S8–17.
67 Hill-Briggs, F., Adler, N. E., Berkowitz, S. A., Chin, M. H., Gary-Webb, T. L., Navas-Acien, A., Thornton, P. L. and Haire-Joshu, D. (2021), 'Social determinants of health and diabetes: a scientific review', *Diabetes Care*, 44, 258–79.
68 Niemuth, N. J. and Klaper, R. D. (2015), 'Emerging wastewater contaminant metformin causes intersex and reduced fecundity in fish', *Chemosphere*, 135, 38–45.
69 Ashworth, J. (2022), 'Drug pollution is threatening the water quality of the world's rivers', Natural History Museum. https://www.nhm.ac.uk/discover/news/2022/july/drug-pollution-threatening-water-quality-worlds-rivers.html [accessed 11 April 2023].
70 Lacobucci, G. (2019), 'NHS prescribed record number of antidepressants last year', *The BMJ: British Medical Journal*. https://www.bmj.com/content/364/bmj.l1508 [accessed 6 May 2024].
71 Burns, C. (2022), 'Antidepressant prescribing increases by 35% in six years', *The Pharmaceutical Journal*. https://pharmaceutical-journal.com/article/news/antidepressant-prescribing-increases-by-35-in-six-years [accessed 11 April 2023].

72 Ford, A. T. and Fong, P. P. (2016), 'The effects of antidepressants appear to be rapid and at environmentally relevant concentrations', *Environmental Toxicology and Chemistry*, 35, 794–8.
73 Bean, T. G., Boxall, A. B. A., Lane, J., Herborn, K. A., Pietravalle, S. and Arnold, K. E. (2014), 'Behavioural and physiological responses of birds to environmentally relevant concentrations of an antidepressant', *Philosophical Transactions of the Royal Society B*, 369, 20130575.
74 University of York (2015), 'Could Prozac be killing off our starlings?' https://www.york.ac.uk/research/themes/prozac-and-starlings/ [accessed 29 April 2023].
75 Ashworth (2022), 'Drug pollution is threatening the water quality'.
76 Ballester, J., Quijal-Zamorano, M., Méndez Turrubiates, R. F., Pegenaute, F., Herrmann, F. R., Robine, J. M., Basagaña, X., Tonne, C. et al. (2023), 'Heat-related mortality in Europe during the summer of 2022', *Nature Medicine*, 29, 1857–66.
77 ONS and UK Health Security Agency (UKHSA) (2022), 'Excess mortality during heat-periods: 1 June to 31 August 2022', London: Office for National Statistics and UK Health Security Agency. https://www.ons.gov.uk/peoplepopulationandcommunity/birthsdeathsandmarriages/deaths/articles/excessmortalityduringheatperiods/englandandwales1juneto31august2022.
78 London Climate Change Partnership, 'Heatwaves'. https://climatelondon.org/climate-change/heatwaves/ [accessed 13 May 2024].
79 Lindley, S. (2022), 'Communities most vulnerable to climate change – background briefing', University of Manchester & Friends of the Earth UK. https://friendsoftheearth.uk/climate/millions-risk-extreme-heat-unless-climate-goals-met [accessed 9 April 2024].
80 Locke, D. H., Hall, B., Grove, J. M., Pickett, S. T. A., Ogden, L. A., Aoki, C., Boone, C. G. and O'Neil-Dunne, J. P. M. (2021), 'Residential housing segregation and urban tree canopy in 37 US cities', *npj Urban Sustainability*, 1, 15.
81 Hoffman, J. S., Shandas, V. and Pendleton, N. (2020), 'The effects of historical housing policies on resident exposure to intra-urban heat: a study of 108 US urban areas', *Climate*, 8, 12.
82 Robinson, J. M., Breed, A. C., Camargo, A., Redvers, N. and Breed, M. F. (2024), 'Biodiversity and human health: a scoping review and examples of underrepresented linkages', *Environmental Research*, 246, 118115.
83 UN Environment Programme World Conservation Monitoring Centre (2021), 'From knowledge to action: Colombia's national ecosystem assessment', UNEP-WCMC. https://www.unep-wcmc.org/en/news/from-knowledge-to-action-colombias-national-ecosystem-assessment [accessed 6 May 2024].

CHAPTER 7

1 The story of the Spix's macaw is charted in my 2002 book: Juniper, T. (2002), *Spix's Macaw: The Race to Save the World's Rarest Bird* (London: Fourth Estate).
2 Chancel, L. (2022), 'Global carbon inequality over 1990–2019', *Nature Sustainability*, 5, 931–8.
3 Kramer, K. and Ware, J. (2019), 'Hunger strike: the climate and food vulnerability index', London: Christian Aid. https://mediacentre.christianaid.org.uk/top-10-hungriest-countries-contribute-just-0-08-of-global-co2-new-report/.
4 Gore, T. (2015), Extreme carbon inequality: why the Paris climate deal must put the poorest, lowest emitting and most vulnerable people first', Oxfam International.

https://policy-practice.oxfam.org/resources/extreme-carbon-inequality-why-the-paris-climate-deal-must-put-the-poorest-lowes-582545/ [accessed 9 April 2024].

5   Chancel, L., Piketty, T., Saez, E. and Zucman, G. (2022), 'World Inequality Report 2022', World Inequality Lab. wir2022.wid.world.
6   Hansen, J. (2008), 'Statement of James E. Hansen in defence of the Kingsnorth Six: testimony for the criminal trial in Kent, United Kingdom.' https://www.columbia.edu/~jeh1/mailings/.
7   de Wit, M., Hoogzaad, J. and von Daniels, C. (2020), 'The Circularity Gap Report 2020', Amsterdam: Circle Economy.
8   United Nations (2019), 'The Sustainable Development Goals Report 2019', New York: United Nations.
9   United Nations (2019), 'The Sustainable Development Goals Report 2019'.
10 Singh, A. K., Kumar, A. and Chandra, R. (2022), 'Environmental pollutants of paper industry wastewater and their toxic effects on human health and ecosystem', *Bioresource Technology Reports*, 20, 101250.
11 Ward, N. (2024), *Horses, Power and Place: A More-Than-Human Geography of Equine Britain*. London: Routledge.
12 Jefferys, P. (2013) Shelter, 'A fair way? Do we prioritise golf or homes?' Shelter. https://blog.shelter.org.uk/2013/11/a-fair-way-do-we-prioritise-golf-or-homes/ [accessed 31 May 2024].
13 Earth Overshoot Day (2018), 'How many Earths? How many countries?', Ecological Footprint Network. https://overshoot.footprintnetwork.org/how-many-earths-or-countries-do-we-need/ [accessed 31 May 2024].
14 The Organisation for Economic Co-operation and Development (OECD) is a grouping of 38 of the world's largest and most developed economies, including European nations, North America, Australia and Japan. Its mission is to promote 'better policies for better lives' across the world.
15 Juniper, T. (2022), *The Science of Our Changing Planet*. London: Dorling Kindersley.
16 United Nations (2022), World Population Prospects 2022: Summary of Results. New York: United Nations Department of Economic and Social Affairs, Population Division.
17 World Population by Year – Worldometer. https://www.worldometers.info/world-population/world-population-by-year/ [accessed 24 March 2023].
18 OECD (2023), 'Catching up with Climate Ambitions', *Global EV Outlook 2023*. Paris: OECD Publications.
19 Maddox, T., Howard, P., Jenner, N. and Knox, J. (2019), Report: 'Forest-smart mining: identifying factors associated with the impacts of large-scale mining on forests'. Washington, DC: World Bank Group. http://documents.worldbank.org/curated/en/104271560321150518/Forest-Smart-Mining-Identifying-Factors-Associated-with-the-Impacts-of-Large-Scale-Mining-on-Forests.
20 Deslandes, N. (2022), 'Almost 400 new mines needed to meet future EV battery demand, data finds', *Tech Informed*. https://techinformed.com/almost-400-new-mines-needed-to-meet-future-ev-battery-demand-data-finds/ [accessed 28 May 2023].
21 Henriques, M. (2023), 'The looming threat of deep-sea mining', *BBC Future*. https://www.bbc.com/future/article/20230310-what-does-the-high-seas-treaty-mean-for-deep-sea-mining [accessed 28 May 2023].
22 Jaskula, B. W. (2021), 'Lithium: mineral commodity summary', US Geological Survey. https://pubs.usgs.gov/periodicals/mcs2021/mcs2021-lithium.pdf.
23 Shine, I. (2022), 'Electric vehicle demand – has the world got enough lithium?' *World Economic Forum*. https://www.weforum.org/agenda/2022/07/electric-vehicles-world-enough-lithium-resources/ [accessed 30 May 2023].

24 United Nations (2024), Press release: 'UN-convened panel to address equity, sustainability and human rights across the value chains of critical energy transition minerals'. New York: UN Secretary-General's Panel on Critical Energy Transition Minerals. https://www.un.org/sites/un2.un.org/files/cetm_panel_launch_press_release-04-2024.pdf.
25 Early, C. (2020), 'The new "gold rush" for green lithium', *BBC Future*. https://www.bbc.com/future/article/20201124-how-geothermal-lithium-could-revolutionise-green-energy [accessed 28 May 2023].
26 Morozzo, P. (2020), 'Electric cars are greener than petrol cars – but they're not perfect'. Greenpeace UK. https://www.greenpeace.org.uk/news/electric-cars-greener-petrol-cars/ [accessed 3 June 2024].
27 Buberger, J., Kersten, A., Kuder, M., Eckerle, R., Weyh, T. and Thiringer, T. (2022), 'Total CO_2-equivalent life-cycle emissions from commercially available passenger cars', *Renewable and Sustainable Energy Reviews*, 159, 112158.
28 Guizar-Coutiño, A., Jones, J. P. G., Balmford, A., Carmenta, R. and Coomes, D. A. (2022), 'A global evaluation of the effectiveness of voluntary REDD+ projects at reducing deforestation and degradation in the moist tropics', *Conservation Biology*, 36, e13970.
29 Greenfield, P. (2023), 'Delta Air Lines faces lawsuit over $1bn carbon neutrality claim', *The Guardian*. https://www.theguardian.com/environment/2023/may/30/delta-air-lines-lawsuit-carbon-neutrality-aoe [accessed 14 May 2024].
30 Guizar-Coutiño et al. (2022), 'A global evaluation'.
31 'The carbon con', (2023), *SourceMaterial*. https://www.source-material.org/vercompanies-carbon-offsetting-claims-inflated-methodologies-flawed/ [accessed 14 May 2024].
32 Gössling, S. and Humpe, A. (2020), 'The global scale, distribution and growth of aviation: implications for climate change', *Global Environmental Change*, 65, 102194.
33 Adera, E. (2022), Adaptation Gap Report 2022: 'Too little, too slow: climate adaptation failure puts world at risk'. Nairobi: United Nations Environment Programme. https://www.unep.org/adaptation-gap-report-2022.
34 United Nations (2023), 'Pledges to the loss and damage fund', United Nations Framework Convention on Climate Change (UNFCCC). https://unfccc.int/process-and-meetings/bodies/funds-and-financial-entities/loss-and-damage-fund-joint-interim-secretariat/pledges-to-the-loss-and-damage-fund [accessed 14 May 2024].

CHAPTER 8

1 Extinction Rebellion UK. https://extinctionrebellion.uk/ [accessed 4 June 2024].
2 Juniper, T. (2024), 'A high five for everyone driving Nature recovery – but we must not rest on our laurels', *Natural England*. https://naturalengland.blog.gov.uk/2024/04/23/a-high-five-for-everyone-driving-nature-recovery-but-we-must-not-rest-on-our-laurels/ [accessed 4 June 2024].
3 Bulman, J. (2018), 'Celebrating 60 years of family spending', Office for National Statistics. https://blog.ons.gov.uk/2018/01/18/celebrating-60-years-of-family-spending/ [accessed 4 June 2024].
4 Burns, F., Eaton, M. A., Balmer, D. E., Banks, A., Caldow, R., Donelan, J. L., Douse, A., Duigan, C. et al. (2020), *State of UK Birds 2020*. RSPB, BTO, WWT, DAERA, JNCC, NatureScot, NE and NRW. https://www.bto.org/our-science/publications/state-uks-birds/state-uks-birds-2020.

5 Rigal, S., Dakos, V., Alonso, H., Auniņš, A., Benkő, Z., Brotons, L., Chodkiewicz, T., Chylarecki, P. et al. (2023), 'Farmland practices are driving bird population decline across Europe', *Proceedings of the National Academy of Sciences*, 120, e2216573120.
6 'Save the Cerrado: our climate depends on it', WWF. https://www.worldwildlife.org/pages/save-the-cerrado-our-climate-depends-on-it [accessed 4 June 2024].
7 Juniper, T. (2018), *Rainforest: Dispatches from Earth's Most Vital Frontlines*. London: Profile Books.
8 Hebebrand, C. and Glauber, J. (2023), 'The Russia–Ukraine war after a year: impacts on fertilizer production, prices, and trade flows', *International Food Policy Research Institute*. https://www.ifpri.org/blog/russia-ukraine-war-after-year-impacts-fertilizer-production-prices-and-trade-flows [accessed 10 June 2024].
9 Tinline, P. (2023), 'Why has it taken so long for stagnant pay to become central to UK politics?' https://www.economicsobservatory.com/why-has-it-taken-so-long-for-stagnant-pay-to-become-central-to-uk-politics [accessed 10 June 2024].
10 Spencer, B. and Yorke, H. (2023), 'Ministers quietly abandon "green crap" as focus shifts to food security', *The Sunday Times*. https://www.thetimes.co.uk/article/ministers-quietly-abandon-green-crap-as-focus-shifts-to-food-security-pzxlggm7p.
11 House of Lords (2022), 'Hungry for change: fixing the failures in food'. Authority of the House of Lords, United Kingdom: Select Committee on Food, Poverty, Health and the Environment. https://publications.parliament.uk/pa/ld5801/ldselect/ldfphe/85/85.pdf.
12 Marmot, M. (2020), 'Health equity in England: the Marmot Review 10 years on', *BMJ*, m693.
13 Berners-Lee, M., Kennelly, C., Watson, R. and Hewitt, C. N. (2018), 'Current global food production is sufficient to meet human nutritional needs in 2050 provided there is radical societal adaptation', *Elementa: Science of the Anthropocene*, 6, 52.
14 Edmonds-Brown, V. (2022), 'From "biologically dead" to chart-toppingly clean: how the Thames made an extraordinary recovery over 60 years', *The Conversation*. http://theconversation.com/from-biologically-dead-to-chart-toppingly-clean-how-the-thames-made-an-extraordinary-recovery-over-60-years-180895 [accessed 19 July 2023].
15 Lorenzo, L. (2019), 'Dead river: an environmental history of the Tyne Improvement Commission 1850–1968CE', *North East Childhoods*. https://northeastchildhoods.org/2019/08/24/dead-river-an-environmental-history-of-the-tyne-improvement-commission-1850-1968ce/ [accessed 19 July 2023].
16 Dunton, J. (2024), '"Insufficient investment" means government will miss clean-water targets, watchdog warns', *Civil Service World*. https://www.civilserviceworld.com/news/article/environment-agency-clean-water-targets-insufficient-investment-oep-warns [accessed 10 June 2024].
17 'State of our rivers' (2021), The Rivers Trust 2021. https://storymaps.arcgis.com/collections/6730f10b64184200b171a57750890643?item=1 [accessed 4 June 2024].
18 'Nutrients', The Solent Forum. https://solentforum.org/services/Information_Hubs/css/Nutrients/ [accessed 10 June 2024].
19 'Downing Street promises "action" on water bills' (2013), *BBC News*. https://www.bbc.co.uk/news/uk-politics-24773950 [accessed 4 June 2024].
20 Ofwat Press Office (2017), 'PN 17/17: Ofwat boss talks of the "decade of falling bills"', Ofwat. https://www.ofwat.gov.uk/pn-1717-ofwat-boss-talks-decade-falling-bills/ [accessed 4 June 2024].
21 Vaughan, V. (2023), 'Water firms urged to save money by diluting climate change plans', *The Times*. https://www.thetimes.co.uk/article/water-firms-urged-to-save-money-by-using-low-climate-change-scenario-clean-it-up-c5clpsbn0 [accessed 10 August 2023].

22 Benjamin, A. (2007), 'Stern: climate change a "market failure"', *The Guardian*. https://www.theguardian.com/environment/2007/nov/29/climatechange.carbonemissions [accessed 4 June 2024].

23 Donkin, A., Marmot, M. and Friends of the Earth (2024), 'Left out in the cold: the hidden health cost of cold homes', London: University College London, Institute of Health Equity. https://www.instituteofhealthequity.org/resources-reports/left-out-in-the-cold-the-hidden-impact-of-cold-homes.

24 Way, R., Ives, M. C., Mealy, P. and Farmer, J. D. (2022), 'Empirically grounded technology forecasts and the energy transition', *Joule*, 6, 2057–82.

25 Jack, S. (2023), 'Oil giant Shell warns cutting production "dangerous"', *BBC News*. https://www.bbc.co.uk/news/business-66108553 [accessed 4 June 2024].

26 World Bank Group (2022), 'Pakistan: flood damages and economic losses over USD 30 billion and reconstruction needs over USD 16 billion – new assessment', World Bank. https://www.worldbank.org/en/news/press-release/2022/10/28/pakistan-flood-damages-and-economic-losses-over-usd-30-billion-and-reconstruction-needs-over-usd-16-billion-new-assessme [accessed 5 June 2024].

27 Shamsuddoha, M. and Chowdhury, R. K. (2009), 'Climate change induced forced migrants: in need of dignified recognition under a new protocol', Bangladesh: Equity and Justice Working Group (EquityBD). https://www.equitybd.net/wp-content/uploads/2015/10/climate-refugee-en-bg.pdf.

28 Rutter, J. (2023), 'Will the Uxbridge by-election prove a crossroads for net zero?' Institute for Government. https://www.instituteforgovernment.org.uk/comment/uxbridge-crossroads-net-zero [accessed 10 June 2024].

29 Seddon, P. (2023), 'Rishi Sunak vows not to add "unnecessary" costs to meet green targets', *BBC News*. https://www.bbc.co.uk/news/uk-politics-66290603 [accessed 6 September 2023].

30 Dajnak, D., Evangelopoulos, D., Kitwiroon, N., Beevers, S. and Walton, H. (2019), 'London health burden of current air pollution and future health benefits of mayoral air quality policies', Imperial College London: http://erg.ic.ac.uk/research/home/resources/ERG_ImperialCollegeLondon_HIA_AQ_LDN_11012021.pdf.

31 Whitty. (2022), Chief Medical Officer's annual report 2022.

32 von Uexkull, N., Rød, E. G. and Svensson, I. (2024), 'Fueling protest? Climate change mitigation, fuel prices and protest onset', *World Development*, 177, 106536.

33 Francis, S. (2023), 'Rishi Sunak delays petrol car ban in major shift on green policies', *BBC News*. https://www.bbc.co.uk/news/uk-politics-66871457 [accessed 7 June 2024].

34 United Nations (2023), '"Humanity has opened the gates to hell" warns Guterres as climate coalition demands action', *UN News*. https://news.un.org/en/story/2023/09/1141082 [accessed 7 June 2024].

35 Residential Landlord (2023), 'Prominent green advocate criticises Sunak's energy efficiency reversal'. https://residentiallandlord.co.uk/prominent-green-advocate-criticises-sunaks-energy-efficiency-reversal/ [accessed 7 June 2024].

36 Oxfam (2022), 'Profiting from pain: the urgency of taxing the rich amid a surge in billionaire wealth and a global cost-of-living crisis', Oxfam International. https://www.oxfam.org/en/research/profiting-pain.

37 Laville, S. (2020), 'England's privatised water firms paid £57bn in dividends since 1991', *The Guardian*. https://www.theguardian.com/environment/2020/jul/01/england-privatised-water-firms-dividends-shareholders [accessed 7 June 2024].

38 Semieniuk, G., Weber, I. M., Weaver, I., Wasner, E., Braun, B., Holden, P., Salas, P., Mercure, J-F. and Edwards, N. R. (2024). Distributional implications and share ownership of record oil and gas profits. https://scholarworks.umass.edu/entities/publication/e17a0088-a106-45f6-a94f-5d8b53dc5b1c.
39 Jack, S. and Edser, N. (2023), 'Shell reports highest profits in 115 years', *BBC News*. https://www.bbc.co.uk/news/uk-64489147 [accessed 7 June 2024].
40 Mathers, M. (2024), 'Shell boss paid £8m in 2023 as oil and gas giant waters down climate pledge', *The Independent*. https://www.independent.co.uk/news/uk/home-news/shell-energy-oil-gas-climate-pledge-b2512445.html [accessed 7 June 2024].
41 Garcia, L. and Stronge, W. (2022), 'A climate fund for climate action: the benefits of taxing extreme carbon emitters', Autonomy. https://autonomy.work/wp-content/uploads/2022/10/A-Climate-Fund-for-Climate-Action.pdf.

CHAPTER 9

1 'Disappearance of the Beothuk' (2013), Newfoundland and Labrador Heritage. https://www.heritage.nf.ca/articles/indigenous/beothuk-disappearance.php [accessed 8 June 2024].
2 Ravalli, R. (2009), 'The near extinction and reemergence of the pacific sea otter, 1850–1938', *The Pacific Northwest Quarterly*, 100(4), 181–91.
3 Ontario Parks, (2022), 'The beaver: architect of biodiversity'. https://www.ontarioparks.ca/parksblog/the-beaver-architect-of-biodiversity/ [accessed 8 June 2024].
4 IMF Research Department (2000), 'The world economy in the twentieth century: striking developments and policy lessons', in: *World Economic Outlook*, May 2000, Chapter 5. International Monetary Fund: IMF.org Publications.
5 Abramovitz, M. (1986), 'Simon Kuznets 1901–1985', *Journal of Economic History*, 46, 241–6.
6 Kuznets, S. (1934), *National Income, 1929–1932*. New York: National Bureau of Economic Research. http://www.nber.org/books/kuzn34-1.
7 Abramovitz, M. (1959), *The Welfare Interpretation of Secular Trends in National Income and Product*. Stanford, CA: Stanford University Press.
8 Kennedy, R. F. (1968), Remarks of Robert F. Kennedy at the University of Kansas. John F. Kennedy Presidential Library and Museum Archive. https://www.jfklibrary.org/learn/about-jfk/the-kennedy-family/robert-f-kennedy/robert-f-kennedy-speeches/remarks-at-the-university-of-kansas-march-18-1968.
9 Cardwell, M. (2013), 'Attenborough: poorer countries are just as concerned about the environment', *The Guardian*. https://www.theguardian.com/environment/2013/oct/16/attenborough-poorer-countries-concerned-environment [accessed 10 June 2024]. The sentiments expressed in this statement were first conveyed by economist Kenneth Boulding in the 1950s and have been repeated by others, including Paul Ehrlich in the 1990s.
10 Keynes J. M. (1936), *The General Theory of Employment, Interest and Money*. London: Palgrave Macmillan.
11 Smith, A. (1776), *An Inquiry into the Nature and Causes of the Wealth of Nations*. London: W. Strahan and T. Cadell.
12 Hope, D. and Limberg, J. (2022), 'The economic consequences of major tax cuts for the rich', *Socio-Economic Review*, 20, 539–59.

13 'Budgets 1979–1992', *BBC News*. https://www.bbc.co.uk/news/special/politics97/budget97/background/bud1979_92.shtml [accessed 11 June 2024].
14 Kleven, H., Landais, C., Muñoz, M. and Stantcheva, S. (2020), 'Taxation and migration: evidence and policy implications', *Journal of Economic Perspectives*, 34, 119–42.
15 Raworth, K. (2017), *Doughnut Economics: Seven Ways to Think Like a 21st-century Economist*. London: Random House.
16 IMF Research Department (2000), 'The world economy in the twentieth century'.

CHAPTER 10

1 'SkyFi's satellite image confirms massive clothes pile in Chile's Atacama Desert' (2022), *SkyFi*. https://www.skyfi.com/blog/skyfis-confirms-massive-clothes-pile-in-chile [accessed 11 June 2024].
2 Arthur, R. (2022), 'Sustainable fashion: communication strategy 2021–2024', United Nations Environment Programme (UNEP).
3 Remy, N., Speelman, E. and Swartz, S. (2016), 'Style that's sustainable: a new fast-fashion formula', *McKinsey Sustainability*, McKinsey & Company. https://www.mckinsey.com/capabilities/sustainability/our-insights/style-thats-sustainable-a-new-fast-fashion-formula#/ [accessed 11 June 2024].
4 Amed, I., Balchandani, A., Beltrami, M., Berg, A., Hedrich, S. and Rölkens, F. (2019), 'The state of fashion 2019', *McKinsey Sustainability*, McKinsey & Company. https://www.mckinsey.com/~/media/mckinsey/industries/retail/our%20insights/fashion%20on%20demand/the-state-of-fashion-2019.pdf.
5 Bulman, J. (2018), 'Celebrating 60 years of family spending', Office for National Statistics. https://blog.ons.gov.uk/2018/01/18/celebrating-60-years-of-family-spending/ [accessed 4 June 2024].
6 United Nations (2019), 'UN launches drive to highlight environmental cost of staying fashionable', *UN News*. https://news.un.org/en/story/2019/03/1035161 [accessed 27 September 2023].
7 Fraser, M., Conde, Á. and Haigh, L. (2024), 'The circularity gap report 2024: a circular economy to live within the safe limits of the planet', Amsterdam: Circle Economy Foundation. https://www.circularity-gap.world/2024.
8 Higgs, K. (2021), 'A brief history of consumer culture', *The MIT Press Reader*. https://thereader.mitpress.mit.edu/a-brief-history-of-consumer-culture/ [accessed 11 June 2024].
9 Ray, A. (2009), 'Hudson's Bay Company', *The Canadian Encyclopedia*. https://www.thecanadianencyclopedia.ca/en/article/hudsons-bay-company [accessed 19 September 2023].
10 'Edward Bernays, "father of public relations" and leader in opinion making, dies at 103' (1995), *New York Times Archive*. https://web.archive.org/web/20180124200210/http://www.nytimes.com/books/98/08/16/specials/bernays-obit.html [accessed 11 June 2024].
11 Cutlip, S. M. (1994), *The Unseen Power: Public Relations, a History*. Hillsdale, NJ: Erlbaum Associates.
12 'Edward L. Bernays Beech-Nut Packing Co – "Edward L. Bernays tells the story of making bacon & eggs all-American Breakfast"', archived from the original on 18 November 2021 (2014), Museum of Public Relations in New York, YouTube. https://www.youtube.com/watch?v=6vFz_FgGvJI.
13 Tye, L. and Bernays, E. L. (2002), *The Father of Spin: Edward L. Bernays & the Birth of Public Relations*. New York: Holt, Owl Books.
14 Nixon, S. (2016), *Hard Sell: Advertising, Affluence and Transatlantic Relations, c. 1951–69*. Manchester: Manchester University Press.

15 Statista Market Forecast (2024), 'Advertising – worldwide', *Statista*. https://www.statista.com/outlook/amo/advertising/worldwide [accessed 14 June 2024].
16 Ridder, M. (2024), 'Coca-Cola Co.: advertising budget 2023', *Statista*. https://www.statista.com/statistics/286526/coca-cola-advertising-spending-worldwide/ [accessed 14 June 2024].
17 Levine, R. S. (2021), 'Childhood caries and hospital admissions in England: a reflection on preventive strategies', *British Dental Journal*, 230, 611–6.
18 Rogers et al. (2024), 'Estimated impact of the UK soft drinks industry levy on childhood hospital admissions for carious tooth extractions: interrupted time series analysis', *BMJ Nutrition*, Prevention & Health. https://nutrition.bmj.com/content/early/2023/10/31/bmjnph-2023-000714 [accessed 22 January 2025].
19 Watt, S. (2024), 'What is the impact of the UK soft drinks industry levy on childhood tooth decay?', *Nature*, EBD, 25, 91-92. https://www.nature.com/articles/s41432-024-01025-3#:~:text=Hospital%20admissions%20for%20decay%2Drelated,year%202022%20to%202023 [accessed 22 January 2025].
20 'Jamie's sugar rush' (2015), https://www.youtube.com/watch?v=7psynBdrZnA [accessed 31 July 2023].
21 The Sun (2024), 'COCA NO-NO Inside 'world's unhealthiest town' where locals guzzle 160L of coke EACH year – even babies are hooked'. https://www.thesun.co.uk/news/29586288/town-mexico-coke-addicts-babies/https://www.nature.com/articles/s41432-024-01025-3#:~:text=Hospital%20admissions%20for%20decay%2Drelated,year%202022%20to%2020234 [accessed 22 January 2025].
22 Encyclopaedia Britannica (2024), 'The Coca-Cola Company: history, products, & facts', *Encyclopaedia Britannica*. https://www.britannica.com/money/The-Coca-Cola-Company [accessed 17 June 2024].
23 Velandia-Morales, A., Rodríguez-Bailón, R. and Martínez, R. (2022), 'Economic inequality increases the preference for status consumption', *Frontiers in Psychology*, 12, 809101.
24 Melita, D., Rodríguez-Bailón, R. and Willis, G. B. (2023), 'Does income inequality increase status anxiety? Not directly, the role of perceived upward and downward mobility.' *British Journal of Social Psychology*, 62, 1453–68.
25 Melita, D., Willis, G. B. and Rodríguez-Bailón, R. (2021), 'Economic inequality increases status anxiety through perceived contextual competitiveness', *Frontiers in Psychology*, 12, 637365.
26 Pybus, K., Power, M., Pickett, K. E. and Wilkinson, R. (2022), 'Income inequality, status consumption and status anxiety: an exploratory review of implications for sustainability and directions for future research', *Social Sciences & Humanities Open*, 6, 100353.
27 Statista Market Forecast. (2023), 'Luxury goods – worldwide', *Statista*. https://www.statista.com/outlook/cmo/luxury-goods/worldwide#revenue [accessed 2 October 2023].
28 Statista Market Forecast. (2023), 'Luxury goods – Hong Kong', *Statista*. https://www.statista.com/outlook/cmo/luxury-goods/hong-kong [accessed 2 October 2023].
29 Daxue Consulting (2023), 'Hong Kong's luxury market losing ground to Mainland China'. https://daxueconsulting.com/hong-kongs-luxury-market/ [accessed 2 October 2023].
30 Wilkinson, R. G. and Pickett, K. E. (2018), *The Inner Level: How More Equal Societies Reduce Stress, Restore Sanity and Improve Everyone's Well-being*. London: Allen Lane, p. 104.
31 Wilkinson and Pickett (2018), *The Inner Level*.
32 Wilkinson, R. G. and Pickett, K. E. (2009), *The Spirit Level: Why More Equal Societies Almost Always Do Better*. London: Allen Lane/Penguin Group UK, Bloomsbury Publishing.
33 Companies Market Cap (2024), 'Companies ranked by Market Cap'. https://companiesmarketcap.com/#google_vignette [accessed 19 July 2024].

34 World Bank Data (2024), 'GDP (current US$)', World Bank Group. https://data.worldbank.org/indicator/NY.GDP.MKTP.CD?most_recent_value_desc=false [accessed 19 July 2024].
35 Korzeniewski, J. (2009), 'The greening of the Oscars continues with the 2010 Toyota Prius', *AutoBlog*. https://www.autoblog.com/2009/02/20/the-greening-of-the-oscars-continues-with-the-2010-toyota-prius/?guce_referrer=aHR0cHM6Ly93d3cuZ29vZ2xlLmNvbS8&guce_referrer_sig=AQAAAMnoPf2jL4iboKdN6LZJlOCz_Dxm tDdewNiZEp81G8AI5f3cnrmUBLfWroe-iJAILGkPJvf37qoLSRC-AWTDbjiS8_VS7 _RMdQjcu6YUIrAFgWbH_gzXCedlGKTseRs8DuHkLao-oSmTM4DgcaWP3lz MPUc9oxbXwtO1B7xhLx10&guccounter=2 [accessed 17 June 2024].
36 Schwab, K., Porter, M. and Sachs, J. (2002), *The Global Competitiveness Report 2001–2002*. Geneva: World Economic Forum. ISBN 0-19-521837-X.
37 Greenwood, D. T. and Holt, R. P. F. (2010), 'Growth, inequality and negative trickle down', *Journal of Economic Issues*, 44, 403–10.
38 Green, F. and Healy, N. (2022), 'How inequality fuels climate change: the climate case for a Green New Deal', *One Earth*, 5, 635–49.
39 Penney, B. (2019), 'The English Indices of Deprivation 2019', Ministry of Housing, Communities and Local Government. Available at: www.gov.uk/MHCLG.
40 Green and Healy (2022), 'How inequality fuels climate change'.
41 Wilkinson and Pickett (2018), *The Inner Level*.

CHAPTER 11

1 *BBC News* (2021), 'COP26: Charles to say "war-like footing" needed'. https://www.bbc.co.uk/news/uk-59115203 [accessed 18 June 2024].
2 Wollburg, P., Hallegatte, S. and Mahler, D. G. (2023), 'The climate implications of ending global poverty', Development Data Group & Climate Change Group, World Bank.
3 Bruckner, B., Hubacek, K., Shan, Y., Zhong, H. and Feng, K. (2022), 'Impacts of poverty alleviation on national and global carbon emissions', *Nature Sustainability*, 5, 311–20.
4 Wilkinson and Pickett (2018), *The Inner Level*, p. 104.
5 Wilkinson and Pickett (2018), *The Inner Level*.
6 Marks, N., Abdallah, S., Simms, A. and Thompson, S. (2019), 'The Happy Planet Index: an index of human well-being and environmental impact'. London: New Economics Foundation, Friends of the Earth. https://policy.friendsoftheearth.uk/sites/default/files/documents/2019-02/happy_planet_index.pdf.
7 Kubiszewski, I., Costanza, R., Franco, C., Lawn, P., Talberth, J., Jackson, T. and Aylmer, C. (2013), 'Beyond GDP: measuring and achieving global genuine progress', *Ecological Economics*, 93, 57–68.
8 Juniper, T. (2013), *What Has Nature Ever Done For Us? How Money Really Does Grow on Trees*. London: Profile Books.
9 Trade Union Congress (2022), 'High Pay Centre and TUC research finds that median FTSE 100 CEO pay increased 39% in 2021'. https://www.tuc.org.uk/news/high-pay-centre-and-tuc-research-finds-median-ftse-100-ceo-pay-increased-39-2021 [accessed 18 June 2024].
10 Office for Budget Responsibility (2024), 'The fiscal implications of tax gaps'. https://obr.uk/box/the-fiscal-implications-of-tax-gaps/ [accessed 18 June 2024].
11 Advani, A., Burgherr, D. and Summers, A. (2023), 'Taxation and migration by the super-rich', *Social Science Research Network (SSRN) Journal*, doi:10.2139/ssrn.4568741.

NOTES 341

12 Friedman, S., Gronwald, V., Summers, A. and Taylor, E. (2024), 'Tax flight? Britain's wealthiest and their attachment to place', International Inequalities Institute, LSE.
13 IMF (2023), 'Climate change: fossil fuel subsidies', International Monetary Fund. https://www.imf.org/en/Topics/climate-change/energy-subsidies [accessed 18 June 2024].
14 See, G. (2024), 'The social cost of carbon is now US$225 per tonne – what this means for Asia', *Eco-Business*. https://www.eco-business.com/news/the-social-cost-of-carbon-is-now-us225-per-tonne-what-this-means-for-asia/?utm_medium=email&_hsmi=301795181&utm_content=301795181&utm_source=hs_email [accessed 18 June 2024].
15 UNEP (2023), 'Global climate litigation report: 2023, status review', Nairobi: Law Division, United Nations Environment Programme. doi:10.59117/20.500.11822/43008.
16 Luciano Lliuya v RWE AG. (2021), Climate Change Litigation Database. https://climatecasechart.com/non-us-case/lliuya-v-rwe-ag/ [accessed 18 June 2024].
17 United Nations (2022), 'UN General Assembly declares access to clean and healthy environment a universal human right', *UN News*. https://news.un.org/en/story/2022/07/1123482 [accessed 18 June 2024].
18 Well-being of Future Generations (Wales) Act 2015 (2015), Future Generations Commissioner for Wales. https://www.futuregenerations.wales/about-us/future-generations-act/ [accessed 18 June 2024].
19 Natural England (2023), 'Natural England unveils new Green Infrastructure Framework', GOVUK. https://www.gov.uk/government/news/natural-england-unveils-new-green-infrastructure-framework [accessed 18 June 2024].
20 University of Bath, The Somer Valley Innovative Pathway Control (IPC) project. University of Bath research portal: https://researchportal.bath.ac.uk/en/projects/the-somer-valley-innovative-pathway-control-ipc-project [accessed 18 June 2024].
21 Wessex Water, AMP6 Environmental Investigations: collaborative public health pilot project. maps.wessexwater.co.uk/webapps/agol/env/investigations/9110.pdf
22 Wessex Water (2023), Innovative Pathway Control Project in the Somer Valley. maps.wessexwater.co.uk/webapps/agol/env/investigations/IPC.pdf
23 Umbach, F. (2024), 'Still waiting for peak coal', *GIS Reports*. https://www.gisreportsonline.com/r/peak-coal/ [accessed 18 June 2024].
24 Copernicus (2023), 'Copernicus: October 2023 – exceptional temperature anomalies; 2023 virtually certain to be warmest year on record', *Copernicus: Europe's Eyes on Earth*. https://climate.copernicus.eu/copernicus-october-2023-exceptional-temperature-anomalies-2023-virtually-certain-be-warmest-year [accessed 18 June 2024].
25 University of Reading (2024), 'Unprecedented ocean heating shows risks of a world 3°C warmer', University of Reading/News. https://www.reading.ac.uk/news/2024/Research-News/Unprecedented-ocean-heating-shows-risks-of-a-world-3C-warmer [accessed 18 June 2024].
26 Met Office (2025). Rise in carbon dioxide off track for limiting global warming to 1.5°C. https://www.metoffice.gov.uk/about-us/news-and-media/media-centre/weather-and-climate-news/2025/rise-in-carbon-dioxide-off-track-for-limiting-global-warming-to-1.5c.

CHAPTER 12

1 'Majority in Britain back "more tax, more spend"' (2022), National Centre for Social Research. https://natcen.ac.uk/news/majority-britain-back-more-tax-more-spend [accessed 6 December 2023].

2   The Sutton Trust (2023), Sutton Trust Cabinet Analysis 2023 (Rishi Sunak). https://www.suttontrust.com/our-research/sutton-trust-cabinet-analysis-2023/ [accessed 14 November 2023].

3   Wikipedia (2024), 'List of prime ministers of the United Kingdom by education'. https://en.wikipedia.org/wiki/List_of_prime_ministers_of_the_United_Kingdom_by_education#List_of_British_prime_ministers_by_education [accessed 24 June 2024].

4   Hsu, P.-H., Li, K. and Pan, Y. (2024), 'The eco gender gap in boardrooms', European Corporate Governance Institute. http://dx.doi.org/10.2139/ssrn.4281479.

5   The Sasakawa Peace Foundation (2020), 'Gender diversity and climate innovation', BloombergNEF and Sasakawa Peace Foundation. https://assets.bbhub.io/professional/sites/24/BNEF-Sasakawa-Peace-Foundaction-Gender-Diversity-and-Climate-Innovation_12012020_FINAL.pdf

6   Norgaard, K. and York, R. (2005), 'Gender equality and state environmentalism', *Gender & Society*, 19, 506–22.

7   Merkle, O. (2022), 'Anti-corruption and gender: the role of women's political participation', Westminster Foundation for Democracy. https://www.wfd.org/what-we-do/resources/anti-corruption-and-gender-role-womens-political-participation.

8   Montacute (2019), 'Elitist Britain' 2019.

9   Uberoi, E. and Carthew, H. (2023), 'Ethnic diversity in politics and public life', UK Parliament, House of Commons. Available via commonslibrary.parliament.uk

10 The Sutton Trust (2024), 'Sutton Trust analysis of Labour cabinet'. https://www.suttontrust.com/news-opinion/all-news-opinion/sutton-trust-analysis-of-labour-cabinet/ [accessed 22 July 2024].

11 Sobolewska, M. (2024), 'Ethnic diversity in politics is the new normal in Britain', UK Election Analysis. https://www.electionanalysis.uk/uk-election-analysis-2024/section-1-truth-lies-and-civic-culture/ethnic-diversity-in-politics-is-the-new-normal-in-britain/ [accessed 22 July 2024].

12 Ledgerwood, E. and Lally, C. (2024), 'Election turnout: why do some people not vote?' *POST*, UK Parliament. https://post.parliament.uk/election-turnout-why-do-some-people-not-vote/ [accessed 24 June 2024].

13 Patel, P. and Valgarðsson, V. (2024), 'Half of us: turnout patterns at the 2024 general election', London: Institute for Public Policy Research. https://www.ippr.org/articles/half-of-us

14 ABC News (December 18th 2024). 'Trump has tapped an unprecedented 13 billionaires for his administration. Here's who they are.', https://abcnews.go.com/US/trump-tapped-unprecedented-13-billionaires-top-administration-roles/story?id=116872968

15 Dhir, R. K., Cattaneo, U., Cabrera Ormaza, M. V., Coronado, H. and Oelz, M. (2020), 'Implementing the ILO Indigenous and Tribal Peoples Convention no. 169: towards an inclusive, sustainable and just future'. Geneva: International Labour Organization.

16 Garnett, S. T., Burgess, N. D., Fa, J. E., Fernández-Llamazares, Á., Molnár, Z., Robinson, C. J., Watson, J. E. M., Zander, K. K. et al. (2018), 'A spatial overview of the global importance of Indigenous lands for conservation', *Nature Sustainability*, 1, 369–74.

17 Brown, S. (2023), 'New online map tracks threats to uncontacted Indigenous peoples in Brazil's Amazon', *Mongabay*. https://news.mongabay.com/2023/09/new-online-map-tracks-threats-to-brazilian-amazons-uncontacted-indigenous-peoples/ [accessed 3 October 2023].

18 Amos-Flom, K. (2023), 'Triumph and turmoil: The Xokleng case and the future of Indigenous land rights in Brazil', *Columbia Journal of Transnational Law*. https://www.jtl.columbia.edu/bulletin-blog/triumph-and-turmoil-the-xokleng-case-and-the-future-of-indigenous-land-rights-in-brazil [accessed 25 June 2024].

Index

ABC 92
aboriginal Australians 63, 306
Abramovitz, Moses 213
Adbusters 244
Adoo-Kissi-Debrah, Ella 84
Adoo-Kissi-Debrah, Rosamund 84
advertising 234–40
Afghanistan 116–17
agriculture *see also* food sector
   and agricultural revolution 6
   in Cambridgeshire 5–6
   environmental damage from 7–8, 152, 174–82
   and the 'Green Revolution' 6, 44
air pollution 83–5, 90–1
air travel 168–9
al-Assad, Bashar 115, 116
al-Assad, Hafez 115
Aldrin, Buzz vii
Amazon Sacred Headwaters Alliance 88
Amess, David 189
Anthropocene epoch 11–12, 27, 31, 68, 153, 226, 229, 310
Apollo 11 landing vii
Armstrong, Neil vii
Arnold, Kathryn 141
artificial intelligence (AI) 273–4
Attenborough, Sir David 127, 214
Attlee, Clement 43

Bailey, Ron 187, 188, 189, 190
Bali Action Plan 170
Bennett, Craig 81, 82
Berlin Mandate 28–9
Bernays, Edward 236
Bezos, Jeff 288
Biden, Joe 31
Big Ask campaign 32
biodiversity
   Bob Watson on 15, 16
   and Earth Summit (1992) 27
   Friends of the Earth campaign on 37
   and Convention on Biological Diversity (CBD) 27, 28, 34
   and Global Biodiversity Framework 300
   and indigenous peoples 300–1, 303
BirdLife International 19–20, 23–4, 146
Black Girls Hike 135–7
Blade Runner (film) 70
Blueprint for Survival, A 160
Bolsonaro, Jair 38–9
Borneo 85–8
Bradgate Park 16–17
Brazil 7, 38–9, 109, 129, 145–6, 302–3
Bretton Woods 213
Bristol 202–3, 206

British Medical Journal 140
Brundtland, Gro Harlem 22
Brundtland Commission 22–3, 159
Brunel, Isambard Kingdom 205
Bush, George 27
Bush, George W. 31, 113
Buy Nothing Day 243–4
Byrd-Hagel Resolution 30

Cabot, John 203
Cambridge Institute for Sustainability Leadership vii
Cameron, David 185
Campaign for Warm Homes 186, 187–90
car pollution 192–4
carbon offsetting 166–8
Cargill family 197
Carson, Rachel 44
Carter, Jimmy 218
Centre for the Understanding of Sustainable Prosperity (CUSP) 220
Chancel, Lucas 149
Charles III, King vii, 32, 303, 305–6
chemical industry 70–2
Churchill, Winston 280
circular economy 8, 14, 166, 256, 264–5, 273–4
classical economics 215–16
climate change *see also* greenhouse gases
   adaptions to 9–10
   blocks to progress on 38–41, 194–6
   and conflict 114–18
   at COP21 32–3, 96, 97
   culture wars in 194–6
   in developing world 100–4, 170–1
   at Earth Summit (1992) 27
   and food production 119
   and gender 104–7
   and Hurricane Katrina 111–14
   from industrialization 4–5
   and inequality 100–14
   IPCC reports on 31–2, 33–4, 35–6, 100
   and Kyoto Protocol 29–31
   and Maldives 95–100
   and natural world 142–5
   and pathogen spread 10–11
   and refugees 114
   and Syrian conflict 114–16
   in urban areas 107–11, 143–4
Climate Change Act (2008) viii, 32, 287
Clinton, Bill 71
Clinton, Hillary 249
clothing 92, 230–2
Club of Rome 160
Cobb, John 260
Cobham, Alex 52
Coca-Cola 238–9
colonialism
   and economic growth 202–10
   and greenhouse gas emissions 162–4
   and resource extraction 206–7
Commission for Racial Justice 74
Conference of the Parties (COP)
   COP1 (Berlin) 28–9
   COP3 (Kyoto) 29
   COP16 (Cali) 300
   COP21 (Paris) 32–3, 96, 97, 195
   COP26 (Glasgow) 36, 170
   COP27 (Sharm El Sheikh) 170–1
   COP28 (Dubai) 38, 104, 171, 305
   COP29 (Baku) 171
consumerism
   and advertising 234–40
   and clothing 230–2
   Friends of the Earth campaigns 243–4

and inequality 245–7
rise of 232–4
and status anxiety 240–7
consumption
  environmental damage from 8–9, 151–4
  incentives for sustainable 274–6
  and inequality 153–4
Convention on Biological Diversity (CBD) 27, 28, 34
Cool Earth 11, 304
Cooper, Leonie 84
Costa Rica 258
Côte d'Ivoire 100–1, 114
COVID-19 10, 50, 51, 55–6, 65–6
Cowley car factory 77
Cox, Jonson 185
Cuba 258

Daly, Herman 260
Davies, Julia 287–9, 296
deforestation 85–90, 100–1
Degai, Tatiana 302
Delta Air Lines 167–8
developing world
  and climate change 100–4, 170–1
  and colonialism 162–4
  and decline in absolute poverty 49–51
  environmental links to 34–5, 39–40
  and environmental protections 155–61
  and globalization 78–83
  and the 'Green Revolution' 6, 44
  loss and damage funding 170–1
  need for greater development 155–61
Diabetes UK 139
Douglas, Michael 48
D'Sa, Desmond 75, 80
Douglas, Oronto 80

Earth Summit (1992) 23–8, 34, 35, 155, 159, 170, 219
Earthrise photo 23
Ecologist, The 160
economic growth
  and classical economics 215–16
  and colonialism 202–10
  in developing world 155–61
  economic damage of 9–10
  green growth 226–9
  and inequality 222–6
  and Keynesianism 215
  measurement of 210–15, 220–2
  and neoliberalism 216–22
  in post-war years 42–4
Ecuador 88–9
education 66–7
Eisenhower, Dwight D. 40–1
electric vehicles 164–5, 166, 169–70
Emissions Gap Report 36
energy efficiency 186–92
Enns, Eli 204, 301
Environment Agency 78
Environmental Protection Agency 73, 93
Equality Trust 58
Evans, Chris 59
Extinction Rebellion 172–3
Exxon Valdez 20

Factory Watch resource 70–8
Farman, Joe 24
Fatinikun, Rhiane 135–7
Fauna & Flora 167, 168
food sector 152, 174–82, 196–7 *see also* agriculture
Fox, Laura 167, 168
Foy, Claire 59
Friedman, Milton 216
Friends of the Earth
  access to the natural world study 134
  and air pollution surveys 83–4

Big Ask campaign 32
biodiversity campaigns 37
and Campaign for Warm
  Homes 186, 187–90
campaigns under Tony
  Juniper 19–22
consumerism campaigns 243–4
at COP21 32
and Earth Summit (1992) 25–6,
  27–8
and energy efficiency
  campaigns 186, 187–90
and Factory Watch resource 70–8
and Kyoto Protocol 29–30
Friends of the Earth
  International 79–82
fuel costs 198–9

Gandhi, Indra 159
gender *see* women
General Theory of Employment, Interest and Money, The (Keynes) 215
generational inequality 68
Genuine Progress Indicator
  (GPI) 260
gilets jaunes movement 193–4
Gilligan, Elaine 76–7
Gini, Corrado 45
Gini coefficient 45–6, 47–9, 52, 53,
  55
Global Biodiversity Framework 300
Global Climate Coalition 29
Global Footprint Network 153, 154
globalization 78–83
golf courses 152–3
Gove, Michael 173
Great Acceleration, The 11–18, 44,
  153
Great Britain, SS 205–6
green growth 226–9
Green New Deal 266, 279
Green New Deal ISA 268
'Green Revolution' 6, 44

greenhouse gases *see also* climate
  change
and air travel 168–9
and carbon offsetting 166–8
and colonialism 162–4
inequalities in production
  of 148–50, 155–8
and poverty 255
Greenpeace Germany 92
Gross Domestic Product
  (GDP) 210–15, 220–2, 227–8,
  229, 257–9
Gummer, John 29
Guterres, António 102, 165–6

Hansen, James 148
Happy Planet Index 259–60
Hayek, Friedrich 217
Her Majesty's Inspectorate of Pollution
  (HMIP) 71, 72
Hurricane Katrina 111–14
Hurth, Victoria 262, 297

Index of Sustainable Economic
  Welfare (ISEW) 260
indigenous peoples 62–3, 85–90,
  203–5, 261, 277, 286, 298–306
industrialization
  and climate change 4–5
  in middlesbrough 1–2
  paradox of 2–3
  and population growth 3–4
inequality
  and access to natural world
    131–8, 285
  and climate change 100–14
  and consumerism 245–7
  and consumption levels 153–4
  decline in post-war years 46–7
  and decline in poverty 51–8
  and economic growth 222–6
  and education 66–7
  and energy costs 198–9

in food sector   196–7
and gender   59–61, 104–7
generational   68
and Gini coefficient   45–6, 47–9, 52, 53, 55
and globalization   78–83
increase in from 1980s   47–9
in just transition   284–5
middle class experiences   51–2
and pollution   71–83, 84
and racial discrimination   61–6, 73–4, 84, 134–8
and trust levels   246–52, 288–9
'Inequality Kills' (Oxfam)   55, 56
Institute of Economic Affairs (IEA)   217
Intergovernmental Panel on Climate Change (IPCC)   14, 15, 17, 26, 29, 31–2, 33–4, 35–6, 100
Intergovernmental Science-Policy Platform on Biodiversity and Ecosystem Services (IPBES)   14
Internal Displacement Monitoring Centre (IDMC)   119
International Fund for Agricultural Development   103
International Labour Organization (ILO)   298
International Monetary Fund (IMF)   210, 213, 270
International Union for Conservation of Nature (IUCN)   204
investments in just transitions   263–6, 289

Jackson, Tim   220–2, 227, 258–9
James, Oliver   247
Joe, Komeok   86
Johnson, Boris   174, 192
Just Stop Oil   277–8
just transition principles
integrated policies   278–80
investments   263–6

laws to protect future generations   278
measurement of progress   257–60
priorities in   256–7
public expenditure   270–1
purpose-led companies   260–3
rights to a clean environment   276–8
sectoral planning   271–4
sustainable consumption   274–6
taxation   266–70

Kelley, Colin   114
Kelley, Hilton   80–1
Kennedy, Robert F.   213, 214
Keruan   86
Keynes, John Maynard   215
Keynesianism   215
Khan, Sadiq   85
Khan, Shahda   309–10
Kuznets, Simon   211–13
Kyoto Protocol   29–31

Laffer curve   268–9
Lang, Tim   179–80
Larson, Denny   74–5
Lean, Geoffrey   24, 25, 28, 159, 160, 219
Lees, Andrew   70–1, 306
Liberia   167–8
Lim, Celine   86–7
Limits to Growth, The (Club of Rome)   160
Lliuya, Luciano   277
loss and damage funding   170–1
Louv, Richard   125
Lost Child in the Woods (Louv)   125

Macmillan, Harold   42, 43, 216
Mad Men (TV series)   234–5
Maldives   95–100
Manhattan Institute for Policy Research   217

Marmot Review 181, 190
May, Theresa 173, 292
media regulation 294–5
middlesbrough 1–2, 308–10
Millennium Development Goals 35
Mind, Brain and Behaviour Research Centre 240
Miss Conduct 206–7, 241
Mont Pelerin Society 217
Montreal Protocol (1987) 24–5
Murphy, Richard 266–8
Musk, Elon 296

National Centre for Social Research 289
National Health Service (NHS) 110, 139
National Income, 1929–1932 (Kuznets) 212
Natural England vii, 135, 172
Natural History Museum 142
natural world
   and climate change 142–5
   connections to 127–31
   and food production 176–8
   health benefits from access to 124–7, 138–9, 285
   pharmaceutical pollution 139–42
   unequal access to 131–8, 285
Nature 9, 149
nature-deficit disorder 125–6
Neale, Greg 28
neoliberalism 216–22
net zero
   legislation for in UK 173–4
   timetable for 15–16
New Orleans 111–14
Niger 117–18
Nixon, Sean 236–8, 240–1, 242, 244–5
Nugee, Richard 116, 117–18

Obama, Barack 31, 113

Office for National Statistics 59, 175
Our Common Future (Brundtland Commission) 22–3, 34, 159
Owen, Matthew 305
Oxfam 51, 55, 56, 148–9, 156
ozone 'hole' 24, 25, 31

Palma, José Gabriel 52
Palma ratio 52–3
Panel on Critical Energy Transition Minerals 165–6
pathogen spread 10–11, 55
Patriotic Millionaires 287–90, 294
Patten, Chris 22
Pearl, Morris 289–90
Peas, Domingo 88–9, 303, 304
Pew Research Center 113
Pharmaceutical Journal, The 140
pharmaceutical pollution 139–42
Pickett, Kate 57, 245–6, 247, 248, 308
Pike, Mary 188
plastic 8, 91–3
political ideologies 285–7
political reform 292–4
pollution
   and air pollution 83–5, 90–1
   and Factory Watch resource 70–8
   from cars 192–4
   and globalization 78–83
   and inequality 71–83, 84
   pharmaceutical pollution 139–42
   in rivers 182–5
population growth 304, 157–8
Potsdam Institute 9
poverty
   decline in absolute 49–51
   and diet 181
   and greenhouse gas emissions 255
   and rise in inequality 51–8
Prelude, The (Wordsworth) 130
Profumo scandal 42
public expenditure 270–1

purpose-led companies 260–3, 290

Qatar 108–9

racial inequalities
   and access to natural world 134–8
   and climate change 111–14
   and pollution 73–4, 84
   varieties of 61–6
   voices of campaigners heard 286
Rainforest (Juniper) 101
Ramblers Association 135
Reagan, Ronald 217, 218, 224
recycling of plastics 8, 91–3
Rees-Mogg, Jacob 195
resource extraction
   and colonialism 206–7
   for electric vehicles 164–5, 166
   environmental impact of 85–90
   greener technologies for 165–6
   and rising consumption 150–1
Richard, Margie 80
Rio Summit *see* Earth Summit
Rockström, Johan 11–12, 14, 68

Sandys, Laura 195
Saro-Wiwa, Ken 80
SAVE Rivers 86, 87, 88, 277
Sawan, Wael 191
Scott, Ridley 70
sectoral planning 271–4
Shauna, Aminath 97–8, 99–100
Shell 79–82, 191–2, 199
Shiva, Vandana 162–3
Silent Spring (Carson) 44
Smith, Adam 215–16, 224
Smith, Matt 59
Smith, Susan 53–4, 150, 225
social capital 247–52
social media 295–7
Social Mobility Commission 66
soil degradation 7

Somalia 102–3
Sommestad, Lena viii
Soros, George 206
South Africa 48–9
species decline 7, 27, 176–8
Spirit Level, The (Pickett & Wilkinson) 57, 245–6
Sriskandarajah, Danny 56
status anxiety 240–7
Stern Review 187
Stockholm Resilience Centre 11
Stone, Oliver 48
Summers, Andy 268–9
Sumner, Andy 52
Sunak, Rishi 192, 194–5, 224
Sustainable Development Goals (SDGs) 35, 41, 52, 56, 103, 264
Sutton Trust 66
Syria 114–16
Szreter, Simon 54, 218–19

taxation 266–70
Taylor, Mary 71, 72
Thatcher, Margaret 217, 224, 292
Theory of the Leisure Class, The (Veblen) 235
This Common Inheritance (Department of the Environment) 21–2
Tolba, Mostafa 25
Trade Union Congress (TUC) 263
Trump, Donald 31, 38, 194, 196, 206, 224, 249, 296
Truss, Liz 224
trust levels 246–52, 288–9
Tura, Hope Esquillo 80, 82

Union Bancaire Privée (UBP) 264–5
United Nations
   and Earth Summit (1992) 23–8
   and poverty 50
   and resource extraction 150–1
   on species decline 7

United Nations Environment
    Programme (UNEP)   23, 25,
    171
United Nations Environment
    Programme World Conservation
    Monitoring Centre (UNEP-
    WCMC)   145
United Nations Framework
    Convention on Climate Change
    (UNFCCC)   27
urban areas   107–11, 123–4, 143–4
Uxbridge and South Ruislip by-
    election (2023)   192

Veblen, Thorstein   235

Wall Street (film)   48
Wallich, Henry   228
Warsaw International Mechanism for
    Loss and Damage   170
waste disposal   90–4
water usage   151–2, 182–6, 197–8
Watson, Bob   14, 15–16
Watts, Sir Philip   82
Waygood, Steve   265–6, 290
Wealth of Nations, The (Smith)   215
Well-being of Future Generations
    (Wales) Act (2015)   278

What Has Nature Ever Done For Us?
    (Juniper)   10
Wilkinson, Richard   57, 245–6, 247,
    248
Winkleman, Claudia   59
women
    climate change impact   104–7
    inequality   59–61
    political representation   291–2
    and rights to a clean
        environment   276–7
    voices of heard   286–7
Women's Institute (WI)   286–7
Wordsworth, Willliam   130
World Bank   50, 51, 55–6, 213, 255
World Economic Forum   56
World Environment Day   23
World Health Organization
    (WHO)   83
World Inequality Report (World
    Bank)   56
'World Scientists' Warning to
    Humanity'   26–7

Yorke, Thom   32

Zagrovic, Carole   76
Zuckerberg, Mark   290–1